Black Dragon

Breaking the Frizzle Frazzle of THE BIG LIE of Climate Change Science

Geraint Hughes

Black Dragon: Breaking the Frizzle Frazzle of THE BIG LIE of Climate Change Science

© 2019 Geraint Hughes
Print ISBN 978-1-949267-00-6
ebook ISBN 978-1-949267-01-3

This publication is designed to provide competent and reliable information regarding the subject matter covered. It is written for educational purposes only. However, it is sold with the understanding that the author and publisher are not engaged in rendering legal, financial, or other professional advice.

STAIRWAY PRESS—APACHE JUNCTION

Cover Design by Guy D. Corp
www.GrafixCorp.com

STAIRWAY≡PRESS

www.StairwayPress.com
1000 West Apache Trail #126
Apache Junction, AZ 88120 USA

Foreword

GLOBAL WARMING THEORY is false and everything the liberal media teaches about greenhouse theory is wrong. Now, the great pushback begins. In this book, scientific-sounding techno-babble mumbo jumbo, which bears no resemblance to reality, is decoded. Lies are unraveled and truth revealed.

The "twaddle talk" about how greenhouses work by radiation or the lies that atmospheres back-warm planets and the ever-ready falsehood that Venus suffers a "Runaway Greenhouse Effect" are all absurd.

This book explains why.

The deceptive experiments of the left are laid bare and true science is taught. *Black Dragon* breaks the code of deceit and teaches how greenhouses really work and how to use real thermal radiation equations to quickly approximate temperatures of simple objects in space.

It also explains how light bulbs work and how this mechanism disproves back radiation theory. It explains why a CO_2 gas planet is cold and an Oxygen gas planet is hot. It shows the real factors at work in the infamous two gas bottle experiment. It explains Venus and goes on to prove oil isn't really a fossil fuel.

Oil is unfairly demonized to force people to pay more. It shows what could be wrong with the often used mathematical proof of back radiation using two plates and why a basic two-dimensional flat plate model is inappropriate. This often used "London's Calling 911" two plate model could be completely wrong itself.

London should put the phone down and stop making crank calls.

This book will change how you think about climate change. This true science should be compulsory teaching in all schools.

The brainwashing of the high tax left must end.

—Geraint Hughes

Contents

Preamble 4

Executive Summary of Illumination 6

Current False Teaching of How a Greenhouse Works—via Radiation 7

How a Greenhouse Actually Works—via Convection 11

Conclusion 16

Executive Summary of Illumination 17

First Part of the Equation—the Absorption Side 18

Solar Constant 19

Second Part of Equation—the Emission Side 20

Conclusion 25

Conclusion 42

Executive Summary of Illumination 45

Little Light Reaches the Surface of Venus 45

So, Why is Venus so Hot? 47

Venus Has High Atmospheric Pressures 47

Or, Turpan Depression, China 49

Venus is Volcanic 50

Venus has a Thin Crust 54

Conclusion 57

Two Gas Planets Comparison 58

Executive Summary of Illumination 58

Example 1—Increasing Temperatures by Reducing IR Emissivity 59

Example 2—Decreasing Temperatures by Reducing Solar Absorptivity
.... 60

Example 3—The Importance of A / E Ratios 61

Conclusion 65

Executive Summary 66

The Temperatures of Three Stand Alone Cubes 67

Rectangular Cuboid with Both Boxes Absorbing Sunlight 71

Rectangular Cuboids with One-Half in the Shade 76

Split Cuboids into Front and Back Cubes Separated by 1mm........... 84

Conclusion 93

Executive Summary 95

A Single Two Sided Flat Plate in Space 98

Two Flat Plates Back-to-Back Seperated by 1mm 99

Adding a Heat Transfer Medium to the Mix—Conductance 106

Convection in the Heat Transference Medium 111

Atmosphere Increases the Surface Area for Emission 116

Atmosphere: More Volume and Surface Area than the Earth 122

Conclusion 125

Oil is not a Fossil Fuel 127

Executive Summary 127

Conclusion 133

Final Conclusion 134

Bibliography and Referencing 138

Chapter One 138

Journals / Articles 138

Websites 138

Chapter Two 139

Chapter Three 139

Chapter 4 140

Books 140

Journals 140

Websites 141

Images 142

Chapter 5 143

Chapter 6 143

Chapter 7 143

Chapter 8 143

Preamble

NEVER BEFORE IN the history of man have so many been so completely deceived in such a sustained and conniving manner by so few for so long.

The so-called science of pretend man-made climate change is nothing but "Frizzle Frazzle". The "twaddle talk" of the "Frizzlers" has been specifically designed for the express purpose of deceiving the masses, so they will pay taxes to fix something that isn't broken.

You will hear this "Frizzle Frazzle" on the TV, in the papers, on the news, on the radio, on the internet, at work, school, college and University. The daily flannel of "Global Warming", "Man Made Climate Change" and "Back Radiation of CO_2," is rinsed all over for good measure. The "BIG LIE" is spread in every country, in every language, in every school, stadium and workplace.

Did you ever get the feeling that you are being swindled and deceived by the Carbon Alarmists, the "Frizzlers". Well, your gut feeling is right, you are being conned, robbed of your wealth and you probably fell for it like countless others, hook line and sinker.

Your life is made harder as you have to make cutbacks, change your car and change how you do things at great expense all the while being made to pay higher taxes yet being offered nothing in return. Whilst the lives of those that profit from your extra efforts are made easier.

They get to drive big cars, fly fancy planes, have humongous salaries, well-financed retirements and play games on the market with complex carbon derivative trades.

You're supposed to feel a nice warm glow as you save the future and looked after your children's children—but you feel cold from the constant emptying of your wallet.

Well, CO_2 isn't a pollutant. It is not a greenhouse gas and let no-one tell you otherwise. All the climate science "Frizzlers" teach is nothing but a "LIE!" A scientific-sounding, bunch of mumbo jumbo techno-babble. It's a fake—designed to deceive, then swindle you.

You may have felt like it's all a big lie, but you just weren't able to put your finger on how it is they are deceiving. The bombardment from all sides is relentless. If you question the perceived wisdom of the "97%" then you're a climate denier and a flat Earther. The "Frizzlers" tell you they can't wait for you to "age out."

The green-washed communists see we-the-people as nothing but animals to be farmed, to provide them with easy living and plush retirements, whilst you slave away grinding out a living, barely making ends meet.

It is all "A BIG LIE", nothing but a scam to con you out of your hard-earned wages. Right from the start of my book, which documents my discoveries as I set out on the path to truth, I reveal how you too can repel the Carbon lie and put an end to the communist, socialist high-tax, one-world-order government agenda which drives this sickening deception of the masses—making energy and everything that stems from it more expensive.

They want to control you, they want to steal from you and now, armed with this new information, you can "FIGHT BACK!"

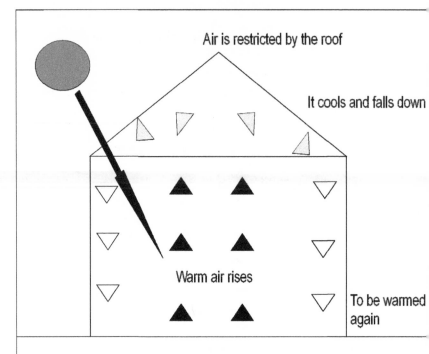

Illumination One—How a Greenhouse Actually Works

Executive Summary of Illumination

- I discuss the current greenhouse theory and how it is taught.
- I show how a greenhouse really works.
- I elaborate on how false teaching is the first step in brainwashing people into believing a lie. My teaching is a necessary step in being able to repel the lies.

Current False Teaching of How a Greenhouse Works—via Radiation

Understanding how the basic tenet of the supposed "Greenhouse Effect" is being mis-taught is an essential first step in understanding how it is that you are being lied to on a global scale. When you know that greenhouses do not work because of "trapped radiation" then you know that you have no need to pay "greenhouse gas taxes".

If you believe a greenhouse gets warmer because of back radiation, then you are wrong and I will explain to you how greenhouses really work. They work due to convection and only convection.

The education establishment teaches the greenhouse effect wrong from first principles and it just goes to show how little they and the Carbon Alarmists—whom all do nothing much other than bang on about the devastating effects of Carbon Dioxide—actually know.

They claim global warming is caused by greenhouse gases and is a man-made threat, This, in turn, drives devastating climate change which will one day wipe us all out. For how many years have they been predicting the end of the world?

I'm actually not sure, it's been so long.

And, the people who spout off about greenhouse gases tend to be the same people who espouse higher taxes for a whole host of other reasons, too. They like to mix it all up together, by making false linkages—such as when a drought happens in one part of the world, they will say "Man-made climate change" in order to make you feel like you did something wrong and now you need to make up for it.

As one example, in the year 2000. David Viner, whom worked for the Climatic Research Unit (CRU) at the University of East Anglia, told the U.K. Independent that "Snowfalls are now just a thing of the past" and that snowfall will be "a very rare and exciting event." [1] This story that has been viral on the internet for a few years, but it goes to show the level to which Climate Alarmists—with their Frizzle Frazzle of nonsense science—will go to deceive people by pretending that CO_2 causes climate change.

Yet, nearly twenty years later, were all still here, the world hasn't ended and we still occasionally get snow in winter.

Their pay packets depend upon acts of deception. They need to publish stories, articles and pretend scientific papers of spurious nonsense or their government-funded grants will dry up.

I put it to you that you should stop paying their wages. They offer nothing in return for the money, so why waste it?

All these scientists, with all their knowledge, their maths, their diagrams and relentless media coverage, how can they possibly be wrong? And more importantly, how do you go about showing to them that they are wrong.

How do you show them up for the Charlatans and frauds that they are?

First, you need to know that Greenhouses DO NOT work because of Back Radiation.

---

[1] UK Independent, 2000

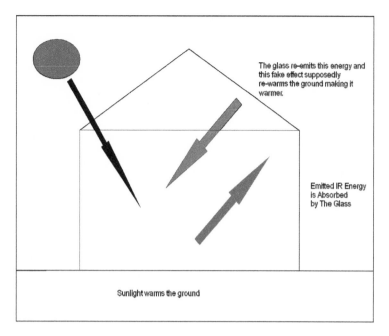

The glass re-emits this energy and this fake effect supposedly re-warms the ground making it warmer.

Emitted IR Energy is Absorbed by The Glass

Sunlight warms the ground

Diagram 1.1

Supposed Greenhouse Effect

You may have seen something like Diagram 1.1; sometimes they use one which shows the atmosphere instead of a greenhouse. It is still wrong.

This backwards drivel is being taught to students worldwide. You should be aware that this is nonsense. As you will realise by end of the next Illumination, the presence of the greenhouse actually exhibits a cooling effect on the surface, because it draws energy through conductance from the ground and then emits that outwardly. Its emissions have a cooling effect on the ground—an effect you will not notice on Earth but one you would notice when you put the greenhouse in space, as I will show in Illumination Two.

The current "completely wrong 97%" consensus theory goes

something along the lines like this:

- Visible solar energy warms up the ground passing unabsorbed through the transparent glass.
- The ground emits infra-red radiation.
- The infra-red radiation is sent out into space.
- The glass blocks this, is warmed by the infra-red, and in becoming warmed, returns a portion of IR back to the surface that otherwise wouldn't have been returned, hence increasing internal temperatures of the greenhouse.
- They then say, this is the "Greenhouse Effect".

There are a multitude of mistakes with this description of events and this multitude of mistakes repeat themselves through the entirety of conventional greenhouse gas theory.

Firstly, it wrongly assumes the glass acts as a source of heat to the ground. It cannot be a source of heat to the ground, because the greenhouse is contact with the ground and it is cooler than the ground. It is being warmed by the ground. Therefore because heat transfers from a hot source to a cold source, and heat transfers from hot to cold.

Secondly, the frame and glass of the greenhouse conducts heat, away from the ground. This can only act to make the ground cooler.

To think that the cold glass is making the ground warmer because of radiation is like saying standing beside a fire, and then saying that you're making the fire warmer, because your skin radiation is going back to the fire.

That's just plain ridiculous.

When I have made this point to "Frizzlers" in the past, they tell me that I don't understand thermal dynamics of radiation. Unfortunately for them, I do. I know it precisely and I know they are wrong because I can do Mechanical Thermodynamics of radiation maths.

And yes, I have had Frizzlers tell me I warm the fire.

Photo 1.1

Is My Hand Warming My Fire?

Is my hand really warming this fire? Or is the fire warming my hand?

How a Greenhouse Actually Works—via Convection

The back radiation theory of greenhouse gas warming was proven wrong in 1909 by Robert R Wood.[2] Professor Wood said:

> *I have always felt some doubt as to whether this action played any very large part in the elevation of temperature. It*

---

[2] *Note on the Theory of Greenhouse*, P319

appeared much more probable that the part played by the glass was the prevention of the escape of the warm air heated by the ground within the enclosure.

Quite often, when you tell alarmists that their greenhouse theory is wrong, they will then follow up that lie with another lie, telling you that "Well, they know that it is wrong, but they are talking in metaphors."

In other words, that they know greenhouse gases don't act like greenhouses, but they don't care and will continue teaching lies anyway. This is a totally non-scientific attitude and an outlook lacking—and tells you a lot about the people whom profess that we should all restrict our CO_2 emissions. And by we, they mean "YOU!"

Why teach something wrong? Why not teach it correctly from the start? I didn't learn how to drive by sitting upside down in the driver's seat. Why do they subject their students to drivel?

Brainwashing, that's why!

Here is the OECD definition of the greenhouse effect.

The greenhouse effect is the warming of the earth's atmosphere caused by a build-up of carbon dioxide and other greenhouse or trace gases that act like a pane of glass in a greenhouse, allowing sunlight to pass through and heat the earth but preventing a counterbalancing loss of heat radiation.[3]

This is a completely wrong definition—and they use this falsehood to drive worldwide governmental policy. This will change; the young people whom have been subjected to these lies are pre-disposed to paying higher taxes and restricting their energy consumption. Also, they are easier to control, which is their aim. I have no trust in such people at all. And I believe that when the peoples of the world realise they have been lied to, they will be

---

[3] OECD, 2017

angry and so they should be.

The plan, from the UN down, is to tax people based on Carbon Dioxide emissions. Here is a quote from the World Bank.

Climate change is one of the greatest global challenges of our time. It threatens to roll back decades of development progress and puts lives, livelihoods, and economic growth at risk.

And...

...means action now. Carbon pricing is an essential part of the solution.[4]

Taxation is the true agenda. This is the driving force behind the worldwide fraud that is the fake science of the "climate change." "Saving the planet" is a front, a pre-text for taking your money.

Fake climate science is all just ONE BIG LIE. Adolf Hitler used this technique and referred to it in his book *Mein Kampf*.

In this they proceeded on the sound principle that the magnitude of a lie always contains a certain factor of credibility, since the great masses of the people in the very bottom of their hearts tend to be corrupted rather than consciously and purposely evil, and that, therefore, in view of the primitive simplicity of their minds they more easily fall a victim to a big lie than to a little one, since they themselves lie in little things, but would be ashamed of lies that were too big.[5]

Adolf Hitler used the big lie to twist stories against the Jews and deceive his people. The lying leaders of the world have adopted Adolf Hitler's sickening approach to brain washing the public and

---

[4] *Why Price Carbon?*, World-bank, 2017
[5] *Mein Kampf*, Adolf Hitler, 1925

have adopted "THE BIG LIE" to repress their populous, fleece them of wealth and force them to buy products they don't need. Tell them to stick it.

This is what is going on. There is a gigantic industry built up around Climate change and the left leaning communist minded liberal Elite classes want to lord it all over you, the plebe peasant. The threat isn't real, it's all just made up and they know it. So should you.

So, how does a greenhouse actually work when it is present within an atmosphere?

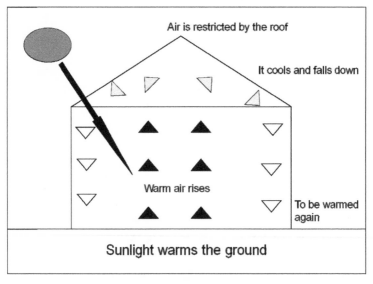

Diagram 1.2

How a Green House Actually Works—Convective Current

Convective Cooling of the Surface causes warming of the air as it contacts the ground. The air inside a greenhouse is subjected to a powerful convection current. This means that air which is touching the surface of the warmed floor has heat conducted to it from the floor, cooling the floor but warming the air. This causes this bottom layer of air to warm, as it warms, its density decreases because gases expand when they are warmed because they have more energy. This causes the air to rise within the greenhouse and cooler, denser air above it to drop.

Just imagine pouring oil and dropping a coin into a bottle of water. The dense coin falls to the bottom but the less dense oil floats on the top. This is what happens inside the greenhouse. This cooler air drops to be in-contact with the floor, which in turn gets warmed, rises and continues the cycle.

The first lot of air, which was warmed, when it falls back to the surface, is warmer than when it first started the cycle, it is warmed again, but to an even higher temperature than before and it is this process of heat cycling which causes the temperature of the air to rise dramatically within greenhouses.

This convection current helps spread warm air around the greenhouse and transfers heat to all the objects within the room, including the external surfaces. These external surfaces prevent the air from escaping the greenhouse.

If there was no roof to the greenhouse, the warm gases would escape and the greenhouse would rapidly cool-down. This is a known method of cooling a greenhouse, you just open up the vents and the inside temperature drops.

This process also occurs outside the greenhouse. However, outside the greenhouse there is no barrier preventing the air from rising, so the air rises miles high and cools much more before it comes back down again. Inside the greenhouse however, the air only rises a short way, before cycling back down. This is why the air inside the greenhouse is much warmer than the air outside a greenhouse.

This is how a greenhouse works.

Thermal radiation from the glass plays no part in warming the greenhouse and as you will see in the next Illumination, thermal radiation emitted from the glass acts to cool the surface, not warm it.

Conclusion

Greenhouses work because of trapped convection currents, not trapped radiation.

Do not let anyone tell you otherwise—they are lying to you if they do. Do not let them teach your children this, they are lying to deceive them, tax and control them.

Most of all, do not trust them, demand to know why you are being lied to and tell your local politician that you're not happy with being made to pay Carbon Taxes when you know the greenhouse effect is a lie.

I hope this puts you on the right path and helps you begin to see how you have been misled by the left-leaning educational establishment.

They do this to disempower people.

They do this to disempower you.

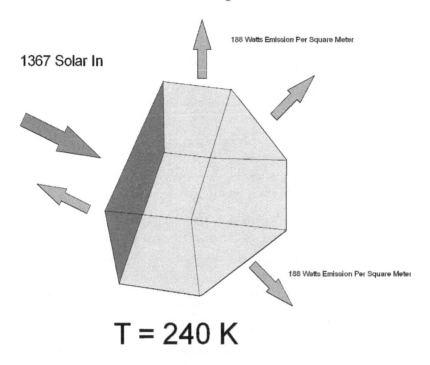

1367 Solar In

188 Watts Emission Per Square Meter

188 Watts Emission Per Square Meter

T = 240 K

Illumination Two—A Flat Plate in Space is Colder with a Greenhouse Attached

Executive Summary of Illumination

- I explain briefly some of the science and maths involved to understand this comparison.
- I show the temperature of a flat plate in space with two sides for emission.
- I show the temperature of a flat plate on the surface of the moon.
- I show the temperature of a cube.
- I show the temperature of a plate with a greenhouse.

The plate with the greenhouse attached, using standard aluminium frame and planes of glass, is colder than just the plate.

First, in order to understand this comparison, I will explain the science involved so you can see how thinking a greenhouse works because of trapped radiation is just plain stupid. I will briefly go over some of the basic information needed to understand this issue.

$$0 = \acute{\alpha} \, E \, A(a) - \acute{\epsilon} \, A \, (e) \, \sigma \, T^4$$

EQ1

Steady State Temperature Equation

This is the equation that describes steady state temperatures of objects in space. We'll consider this before we look at the specifics of how thermal energy is transported. This equation is the first stage of determining temperatures of objects in space and it is an approximate approach. It assumes Isothermal conditions will be achieved and assumes the presence of a vacuum.

It shows that the energy in must balance with the energy out, the principle of energy conservation. The zero tells us that the terms of the equation must balance. This makes sense because we know that any energy that goes into something, must also be sent back out once the object has finished warming up and has reached its steady state temperatures. This is the law of energy conservation.

The first part of the equation is the absorptivity side of the equation. It tells us the amount of solar radiation which is incident upon an object is absorbed by that object.

First Part of the Equation—the Absorption Side

$\acute{\alpha} \, E \, A(a)$. This part of the equation lets us know how much energy is being absorbed by the object. It can be modified to take account of incidence angles, but here we assume maximal, perpendicular conditions.

$\acute{a}$ = Solar Absorptivity of the object, in this case a flat plate, which the unity condition—a black body object. Absorptivity is expressed as a ratio of absorption, where 1 is full absorption and 0 is none. If a surface reflected 10% of incoming radiation, then an absorption ratio would be 0.90. The wavelength range for determining Solar Absorptivity is usually 0-2,500 nanometres at Earth orbit.

E = Solar constant emission value of 1367 W/m^2

Solar Constant

The output of the sun, when expressed in energy terms, is between 64,169,058 W/m^2 and 73,490,000 W/m^2 depending upon what is assumed about the Sun's surface temperature. Some sources say 5,800K while others say 6,000K. This gives a radiant heat output 73.4 Million W/m^2.

When the strongly reductive effect of the Inverse Square Law is applied to radiation, we find we arrive at a Solar Constant of 1,367 W/m^2 at the edge of the Earth's atmosphere. It varies a little bit depending upon the position of the Earth during a year, but this is a generally agreed to average value.

A (a) = Area of absorption, which is the area of the flat plate. We will assume a flat plate is directly perpendicular with the Sun's rays and that the plate is 1 square meter in area. This is to make the numbers easy to follow.

On the energy in side, we see:

$\acute{a}$ E A(a) = 1 (black body plate) x 1367 w/m^2 x 1m^2

The amount of energy absorbed by the flat plate is quite simply 1367 w/m^2. A black body object achieves full absorption of the energy available: 1367 watts. Simple.

The law of energy conversation means Energy In must Equal Energy out. We know energy out must be 1367 watts, but we don't know what temperature the flat plate must achieve to accomplish this, so we do some back-engineering to figure that out.

Second Part of Equation—the Emission Side

The second part of the equation $\acute{\epsilon}$ A (e) σ T^4 is the emissivity side of the equation. This tells us the amount of radiation which is being emitted by an object at a certain temperature.[6]

If the amount of energy being absorbed by an object equals the amount of energy being emitted by an object at steady state conditions, then using these two total figures will enable us to figure out what the temperature of the object is.

$'E$ = Infra-red emissivity of the surface. The emissivity of the plate is considered to be 1, a perfect black body. Again, like absorptivity, emissivity is expressed as a ratio where 1 is full emission across the spectra and 0 is none. The wavelength range for determining infra-red emissivity's is normally 2,500-60,000 nanometres.

A (e) = Area of emission, which is the area of a flat plate at 1 m^2 = 2 square meters if a plate has two sides exposed sides.

σ = Stefan Boltzmann Constant = 5.670367 x 10^{-8} W / (m^2K^4)

T = Temperature of Emissive body at steady state conditions.

We are aiming to find T so the equation needs to be re-arranged to enable us to do this.

EQ 1 : 0 = $\acute{\alpha}$ E A(a) – $\acute{\epsilon}$ A σ T^4

EQ 2 : $\acute{\epsilon}$ A σ T^4 = $\acute{\alpha}$ E A(a)

EQ 3 : T = (A (a) / A * $\acute{\alpha}$E / $\acute{\epsilon}$ σ) $^{1/4}$

EQ 4 : T =($(1m^2)$/ $2m^2$)*((1.0*1367 W/ m^2) / (1.0 * 5.67 x 10^{-8} W / (m^2K))) $^{1/4}$

---

[6] Danish Space Research Institute, 2001

EQ 5 : T = (½ * (1367 / 5.67 x 10^{-8} K)^{¼}
EQ 6 : T = (0.50 * 24,198,347,443)^{¼}
EQ 7 : T = 12,054,673,721^{¼}
EQ 8 : T = 331K or 58 C°

So, what we have is the temperature of a two-sided flat plate at 331K. The freezing point of water or Zero Degrees Celsius is the equivalent of 273K. That means the plate is hot at 58 °C.

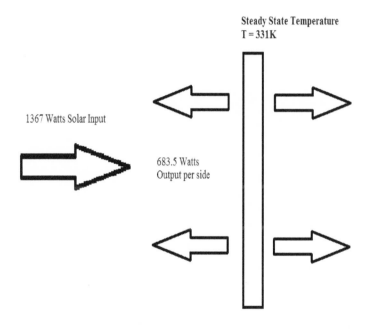

Steady State Temperature
T = 331K

1367 Watts Solar Input

683.5 Watts
Output per side

Diagram 2.1

A Two-Sided Flat Plate in Space

What does this equation tell us?

If the absorptivity and emissivity remain the same, i.e. 1, the only factors which will affect emissive temperature are the areas. So, if we increase the area of absorption, the temperature can increase and if we increase the area of emission, the temperature will decrease. The type of materials used can affect temperatures—as can things like insulation and how an object is assembled. However, there is no need to complicate things—I am expressing the physics from first principles.

Some brief examples will follow.

If we had one-sided plate, it would be much hotter. For example, the Moon's surface is miles deep, so there's no radiation on other side for the plate to radiate out to. The plate can only emit on one side, so it's warmer.

T= 394 K. or 121 °C.

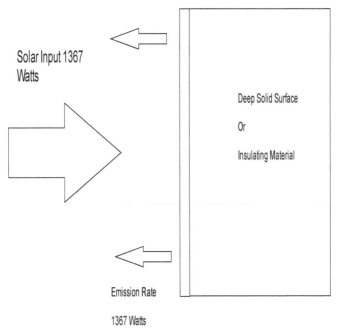

Steady State Temperature of Plate (Ignoring Conductivity to Insulation
T = 394 K

Solar Input 1367 Watts

Deep Solid Surface

Or

Insulating Material

Emission Rate

1367 Watts

Diagram 2.2

A Well Insulated One Sided Plate

We can see that the insulation or deep surface has had a big effect on increasing temperature by 63°C.

A perpendicular cube with 1 side absorbing and 6 sides emitting, is cooler. T = 252 K. This is below zero at -21°C and would therefore be freezing cold to the touch.

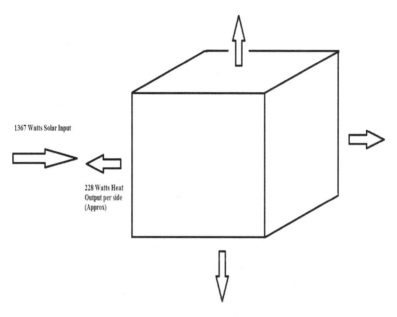

1367 Watts Solar Input

228 Watts Heat
Output per side
(Approx)

Diagram 2.3

Hollow Cube

If we increase the emissive area, the temperature will decrease and if we reduce the emissive area, the temperature will increase, all other things being equal. Again, as a reminder, this approach is an approximate one, works in a vacuum and assumes Isothermal conditions will be achieved and ignores the thermal conductivity of the heated material.

What happens if we stick a greenhouse onto a flat plate in space? Simple, the temperature lowers, because the "energy in" side remains the same, but the "energy out" side has more surface area, so an equivalent amount of energy is now spread across more material. This material vibrates less because the energy is more spread out, resulting in lower temperatures.

Let's say the greenhouse had 1 metre tall sides. (4 sides = 4m^2) and a roof with (2 sides) and a total area of 3m^2. That's a total surface area of 7 m^2. We need to add the side of emission, to get a total area of 8m^2. Let's assume the emissivity is that of glass, 0.90.

We will end up with a temperature of approximately 240K or -33K. This value is less than the cube, despite the glass having a lower emissivity than a black body.

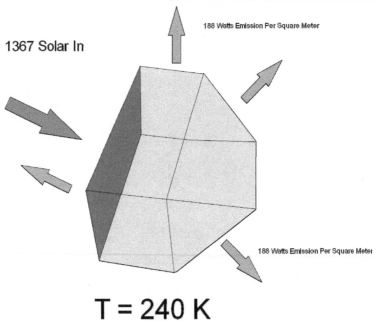

1367 Solar In

188 Watts Emission Per Square Meter

188 Watts Emission Per Square Meter

T = 240 K

Diagram 2.4

Metal Plate with a Greenhouse Attached

The greenhouse has the lowest temperatures because it has the largest areas. To keep this warm in space we would need to add a heating system. The heat from the plate is able to transfer via thermal conductance up the Aluminium frame and then across the glass planes, both of which will radiate that energy out into space. The plate in space all by itself is warmer, because it has less material to transfer heat to. If we added air to the greenhouse, this would add an additional heat transfer path from the warm plate to the frame and panes of the greenhouse and make it more likely that Isothermic temperatures are maintained.

I know what you're thinking. Perhaps if we turned the greenhouse around to face the Sun it would be warmer? No, if we turned the greenhouse by 180°, the temperature would be lower still, because glass is slightly reflective—which means not all of the energy available for absorption would be absorbed. Less energy would be present on the in-side of the equation.

If we assumed 10% reflectivity, then energy in would only be 1230 watts resulting in even lower temperatures as less energy is available to be spread around the greenhouse.

Conclusion

A flat plate in space is warmer $T = 331K$ than a plate with a greenhouse on it where $T = 240K$.

When someone says Greenhouse effect causes warming. You know now, the correct response would be "WHAT ARE YOU TALKING ABOUT?" And when someone talks about Greenhouse Gases you now know to say, "THERE ARE NO SUCH GASES."

There are no unconstrained gases that replicate the physical barrier of a roof of a greenhouse which causes the convective restriction of an actual greenhouse.

FAKE CLIMATE CHANGE

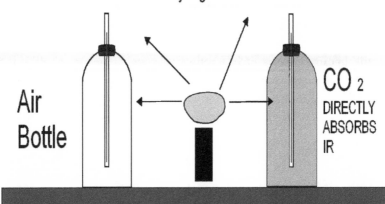

High Quantities of Near and Far IR Emitted By Light Bulb

Air Bottle

CO_2 DIRECTLY ABSORBS IR

GO BACK TO SCHOOL!

Illumination Three—The Deliberately False Experiment

This is a type of experiment often conducted by those whom teach greenhouse gas theory. It's just such a shame that this experiment is nothing but a sham.

The idea is that you demonstrate to observers that the

presence of Carbon Dioxide in a bottle makes it warmer than another equally sized bottle containing just ordinary air from the atmosphere when you shine a light onto the bottles, showing greenhouse gas theory to be real.

Unfortunately, this experiment is wrong and flawed in so many ways, that frankly it's unbelievable. It is nothing but a public act of gross scientific incompetence. Perhaps you know someone whom has performed this experiment for you and failed to inform you of the knowledge I am about to provide. Maybe you will have some pointed questions to ask them relating to why they deceived you so?

If you are unfamiliar with this experiment, you can conduct a search on YouTube or a search engine. Type in "greenhouse effect bottle experiment." You will see many examples of this type of bottle experiment.

This experiment is part of the "BIG LIE" and Climate Change "Frizzlers" will use it to blind you with the techno babble "Frizzle Frazzle". Don't be deceived. Don't let their Carbon Climate Clap Trap get to you.

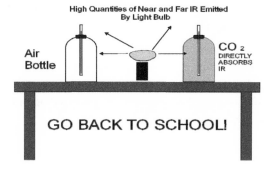

FAKE CLIMATE CHANGE

High Quantities of Near and Far IR Emitted By Light Bulb

Air Bottle

CO_2 DIRECTLY ABSORBS IR

GO BACK TO SCHOOL!

Diagram 3.1

Climate Change in a Bottle

Did you watch any of the videos? Did you find them convincing? Every claim is complete rubbish. The people who do these experiments need to go "Back to School." They don't know what they are talking about and if they do, then they are deliberately misleading everyone with these types of experiments.

They are all wrong for the same reasons. I have seen this experiment done in different ways, some use old bottles, some use mini-greenhouses, some even use large scale greenhouses and add lots of thermometers and insert ice statues and pools of water to spruce things up a little. They are all wrong.

First we must be aware that there are many other factors in determining what temperature a gas will be without looking at its infra-red absorptivity. Without knowing what these other factors you are left in the dark about many scientific facts.

Those other factors which affect this experiment are as follows:

1. Light bulbs emit short-wave IR as well as long-wave radiation and Carbon Dioxide absorbs directly this IR.
2. The bottles aren't sealed correctly.
3. The density of Air and the density of Carbon Dioxide differ.
4. The amount of water vapour in the bottles could be different.
5. The specific heat capacity of Air and the specific heat capacity of Carbon Dioxide are different.
6. The positioning of the light bulbs.
7. The positioning of the temperature probes.
8. Manufacturers deliberately add gases to light bulbs—this can make them cooler and less bright.

These factors cannot be ignored because they have a very real affect on the outcomes of the temperature of the gases involved. These factors are deliberately and unfairly manipulated to give results in favour of the emissive gas over the un-emissive air.

Let's go through each mistake one-by-one and see what the problems are—and what affect they have on the results.

1. Light bulbs emit both Shortwave and Long wave Infrared.

My, oh my, oh my. This is one terrible a mistake to make. This shows the level of unprofessionalism of the people who do this experiment. Did you ever hear them mention this?

A typical tungsten filament light bulb has a surface temperature of the filament of 2800K. What this means is that light bulbs are actually emitting not only light radiation which we can all see, but they are also emitting heat in the form of both Shortwave and Long-wave infra-red radiation.

Most manufacturers will actually state that the light bulbs will emit some 80 to 90% of the inputted energy as IR heat, older style bulbs even more.

This heat is "DIRECTLY ABSORBED" by the IR gases and not by air, which is transparent to IR. The CO_2 gas is absorbing radiation being emitted by the bulb, which bears little to no resemblance to the spectrum of wave lengths being emitted by the surface of the Earth as you can see in Figure 3.2, with earth being closer to the 300K line and the light bulbs being the 3000K and 1000K lines.

The main principle of greenhouse gas theory is that the incoming radiation is absorbed by the "surface" of the Earth, which gets warmed and then in turn emits long-wave infra-red radiation out into space.

The presence of greenhouse gases inhibits this emission out into space and causes radiation energy exchange with the surface, which raises its internal energy levels which then in turn causes the temperature of the surface to rise.

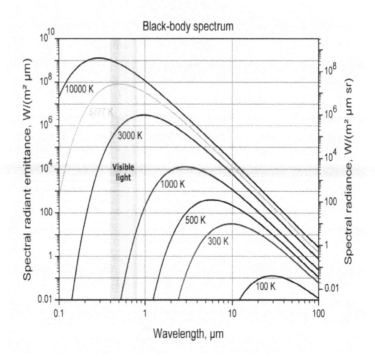

Diagram 3.2

Blackbody Emission Spectrum at Different Temperatures[7]

The experiments "do not" show this, they show the radiation being emitted by the light bulb being absorbed by the CO_2 gas in the containers directly. In other words, BEFORE it reaches the surface. Plastic polyethylene bottles are mostly transparent to IR.

This means we are proving that CO_2 absorbs energy emitted from a light bulb better than air. So what? The Earth isn't heated by light bulbs, it's heated by the Sun, which is millions of miles away, and I hear no one arguing that IR from the Sun is present in such high quantities at the Earth orbit that CO_2 in the air causes direct heating of the atmosphere. Such arguments would fail, because the sun emits far more shortwave UV and Light than it does IR wavelengths applicable to CO_2.

No one disputes that CO_2 absorbs IR, it does. The point is, if you have an emissive gas, and you give it energy by whatever means, along with other ways of dissipating energy, it can radiate, whereas a non-emissive gas doesn't radiate much and generally dissipates its energy via other means, like mechanical collision.

Every climate book I have ever read states quite clearly that CO_2 lets the sunlight through and this energy warms the surface of the earth and CO_2 supposed warming effect, is supposed to be based on inhibiting the release of that energy back out into space.

This bottle experiment and all like it, do no such thing. They warm gas in the bottle directly. It is therefore a completely irrelevant experiment as it doesn't replicate the non-existent greenhouse effect.

Any scientist who doesn't know this is completely incompetent and their teaching is worthless. Ask yourself this question, "Why aren't they telling you this?"

Did you hear the scientists explain direct absorption to you? Why not? The next time you see one of these demonstrations, you know what to ask them.

2. Bottles aren't sealed.

Why is this a problem? When a gas is heated, it starts convection currents moving. This is because the molecules in the gas have more energy, and when the molecules in the gas have more energy they move more vigorously. As they become more vigorous in their movements, that part of the gas becomes less dense than the gas above it. This causes the warmer gas to rise and the cooler gas to fall because the warmer gas literally pushes past the cooler.

If the bottle isn't sealed, the warm gas will rise out of the bottle and will be replaced by cool air from outside the bottle.

If both bottles aren't sealed, this effect will be duplicated by both bottles. However, if one bottle is more open than the other they will experience different rates of leakage.

If you ever see this experiment and you don't see any control measures to prevent leakage, then you know it is unsafe to say that both bottles experienced the same levels of leakage. It is safe to say that the rates of leakage from each bottle will vary.

Variance in leakage rate causes these experiments to be flawed and yield untrustworthy results.

3. The density of air and the density of Carbon Dioxide.

Air has a density of approximately 1.205 kg/m^3. Carbon Dioxide has a density of 1.842 kg/m^3 at room temperature of 20°C.

As I mentioned earlier, when a gas is heated it starts a convection current. This convection current occurs because the density of the warmed gas is less than the cooler gas above it. The warm gas will rise and the cool gas will fall.

When you have a mixture of gases, the heavier gas will fall to the bottom and the lighter gas will rise to the top, even if it is warmer. So, in the bottle with the added Carbon Dioxide, the CO_2 will naturally settle at the bottom and middle of the bottle and the air will naturally rise to the top.

In the bottle of just pure air, the warm gas will rise and the cool gas will fall and this will cause a good deal of mixing. As the gas mixes, its energy will transfer from molecule to molecule. This causes a more even temperature to be found throughout the bottle and will result in lower temperature readings at the probe.

In order for a convection current to get going in the bottle with the added Carbon Dioxide, it must be warmed to a sufficient degree to replace the air above it. It won't do this until it is lighter than the air. "It is necessary therefore for the Carbon Dioxide gas to be much hotter than the air" above it, for its density to be below 1.205 kg/m^3 it would have to rise to a temperature of approximately 170 C°.

In the confines of a plastic bottle being heated by a simple light bulb this simply isn't going to happen, therefore a convection current isn't going to be as strong in the Carbon Dioxide bottle, where the heavier gas will mostly sit at the bottom of the bottle. Only small convection currents within the Carbon Dioxide gas itself could be expected, with little if any mixing between the Carbon Dioxide and the air.

With less movement occurring, there will be less heat transfer from molecule to molecule, so the molecules at the bottom of the bottle will experience more warming. This results in more of the warm Carbon Dioxide gas being at bottom of the bottle.

Also, if the bottles aren't sealed correctly, the pure air bottle will experience more leakage than the Carbon Dioxide mixture bottle, because in the pure air bottle the warm air inside the bottle will be lighter than the cool air outside the bottle and so the warm air will escape from the bottle, cooling the bottled gas.

In the CO_2 bottle, the warm Carbon Dioxide mixture will be heavier than the cool air outside the bottle and won't have a tendency to escape the bottle. The gas in the bottle will experience less cooling, it will be warmer.

The higher density of CO_2 skews this type of experiment in favour of the Carbon Dioxide mixture being warmer than the pure air bottle.

This can be confirmed by using other gases which have densities heavier than air—such as Argon as a replacement for the Carbon Dioxide in the mixed bottle. Argon isn't even considered to be a greenhouse gas and yet a similar warming effect will be seen due to a lack of movement caused by the denser gas in the mixture.

People who perform these experiments should always mention the densities of their chosen gases.

In the table below I have indicated some of the densities of different gases.

Name of Gas	Chemical Symbol	Density in kg/m^3
Air	N / A	1.205
Argon	Ar	1.661
Carbon Dioxide	CO_2	1.842
Chlorine	Cl_2	2.994
Methane	CH_4	0.668

Table 3.1

Density of Gases[7]

We see from these examples that Argon has a density similar to that of Carbon Dioxide and Chlorine has a density which is even greater than Carbon Dioxide, yet neither Argon or Chlorine are considered as greenhouse gases. If this experiment was replicated in the same way that it was here you would find that the temperatures of these gases would be greater than found in just the air bottle as a result of their greater densities.

Did you hear the scientists performing this experiment mention density? No!

Did you hear them tell you that performing this experiment with non-greenhouses gases Chlorine and Argon would yield similar results? No!

Do you think the fact they are omitting this information from, you a bit misleading? You would be right to think so.

4. The amount of water in the bottles is variable.

The bottles used tend to be emptied water bottles and some of the water is bound therefore still to be present on the inside. Without thorough drying, the bottles will be unsuitable for the experiment. If there is no mention of drying procedures to ensure fairness the results can only be untrustworthy.

---

[7] Engineering Toolbox, 2017

Black Dragon

Water is supposed to be a powerful greenhouse gas in its own right and its presence will affect the experiment, contamination prevention measures need to be in place to ensure fairness.

5. The specific heat capacity of air and the specific heat capacity of Carbon Dioxide.

All scientists whom are involved in climatology are aware of something called the specific heat capacity. This is normally measured in KJ / kg / C° and indicates how much energy is required to be input into a substance or gas with a mass of one kilogram to raise its temperature by one degree Celsius.

So, let's say for example air has a specific heat capacity of 1 KJ / kg / C° and another substance, say water, had a SHC of 4 KJ / kg / C°. It would take four times as much energy to raise the temperature of water by 1 degree than it would to cause the air to rise by just one degree. So let's say that 4 kilowatts of heat was inputted into both substances. The air would rise by 4 degrees, but the water would only rise by 1 degree, though both substances received and absorbed exactly the same amount of energy.

Carbon Dioxide Gas has a specific heat capacity of 0.844 KJ / kg / $^\circ$C. This means that for every 1 kilowatt of energy inputted into it, its temperature will rise by 1.18 $^\circ$C, whereas air will rise by only 1 $^\circ$C. This means that Carbon Dioxide can attain much higher temperatures than air for the same input of energy.

However, this also means that it cools down faster. Its temperature fluctuates more than air does. For example, if both bottles were sealed and left to cool outside at night, the CO_2 mixture bottle would cool down faster and experience lower temperatures than the pure air mixed bottle.

If you were not aware of this, you could be misled into believing it is the Carbon Dioxide's ability to absorb infra-red radiation which caused its temperature to rise more than the pure air and not because of its different SHC.

Again, if you used other gases instead of Carbon Dioxide,

different temperatures would be experienced depending upon their specific heat capacities. This is an important point to make, because the people whom perform this experiment whom do not mention this could be accused of deliberately misleading people and rightly so, they should know better.

Name of Gas	Chemical Symbol	Specific Heat Capacity KJ/kg/k
Air	N/A	1.01
Argon	Ar	0.52
Carbon Dioxide	CO_2	0.84
Chlorine	Cl_2	0.48
Methane	CH_4	2.22

Table 3.2

Specific Heat Capacity of Gases[8]

The specific heat capacity should always be mentioned in these experiments.

You can see that if we swapped the CO_2 for Argon or Chlorine, both of which are not considered to be greenhouses gases, much higher warming would be experienced. This is because they have much lower SHC's and so therefore increase in temperatures more rapidly. But there's no need to mention this is there, because this would stop people paying Carbon Taxes and we can't have that now can we?

This factor on its own is powerful enough to alter the temperature of the gases all on its own. Yet never have I heard a "Frizzler" mention it at all when conducting these fake experiments. Never, not even the once over all the years or in any of the demonstrations performed by anyone, anywhere. How disgraceful is that? Just what is their excuse? Have they never heard of this basic fact? How did they get to be professors?

---

[8] Ibid.

6. The positioning of the light bulbs.

I will say it again, because it an important point to make, light bulbs shouldn't even be used. Neither should electric coil heaters or anything with a seriously high temperature output which bears no resemblance to Earth-like temperatures. An adjustable heating mantle would be a much more appropriate alternative, although the results it would give wouldn't be as spectacular as when using a light bulb. Also, when conducting this experiment, cooling rates should be also shown.

The positioning of the light bulbs influences the outcome of the experiment.

If the bulbs aren't exactly the same distances away from the bottles and shining on the bottles in exactly the same location then the temperatures will differ across the bottles, even if they both contained just air. This is because if the light is being received lower down on a bottle it will have a greater heating effect, as the bench will help heat up the bottle. Or if some of the light isn't shining on the bottle, but is instead shining away from the bottle then some heat won't be received by the bottle and it will be cooler as a result. If one of the lamps is even 1cm closer this too will affect the result.

Every time I see these differences, for some reason, I can't think why? They are always wrong to give maximal readings for CO_2 and minimal readings for air. Many times when you witness these experiments you will see that the lamps are positioned differently and are shining on the bottles differently. This causes error to the experiment and means all results are flawed.

Lamp control measures are vital, if they aren't in place the experiment is worthless.

7. The positions of the temperature probes.

The positioning of the temperature probes affects the outcome because if they are in different locations in the bottle, they will experience different readings as a result of the natural stratification process which occurs within heated containers.

For example as warm air tends to rise, if the probe was at the top of the bottle then it would pick up different readings than if the probe was at the bottom. Also, if the probe was spread across the bottle it would experience a higher temperature because more of the light would directly be shining on to the probe, causing it to be warmer.

Many times when you see these experiments, the probes will be in different sections of the bottle, or not in line, or spread diagonally across on one bottle—but in another they will be to one side. They should always be the same on all bottles. This is deliberate, having the CO_2 side positioned to maximise readings and the air sensor positioned to minimise readings.

To perform this experiment correctly the probes should be inserted in through the top of the container, exactly in the centre of the top, facing straight down into the bottle, with no angle on the probe at all, with all bottles exactly the same.

8. Manufacturers deliberately add gases to light bulbs—this can make them cooler and less bright.

A modern day light bulb works by passing an electric current through a tungsten filament. This heats up the filament which—when it reaches a high temperature—it will glow red or white hot. The heat causes the filament to evaporate and thin over time as the high heat boils molecules of the filament off. This causes the lifespan of the bulb to reduce and as the wire gets thinner, it gets hotter which then further reduces the lifespan.

To overcome this effect, manufacturers can add a gas to the bulb. This has the effect of reducing the temperature of the wire as heat energy is transferred away from the wire via conduction and convection. This transfer of heat causes the wire to cool and glow less brightly, but as the gas heats up it can cause the bulb surface to be hotter than if the gas wasn't present, as energy is transferred more rapidly from the wire to the glass bulb with the gas present than if the gas was present.

Oxygen is a highly reactive gas, so air which contains oxygen isn't used as this would burn the filament more quickly, so inert gases such as helium, argon, neon and xenon are used.

The point is that the conductive and convective effects of these gases are well known in industry to cause the temperature of the filament to drop and glow less brightly. This can be seen in the picture below taken from the website indicated.

Diagram 3.3

Light Bulb with a Vacuum & Gas[9]

---

[9] http://www.lamptech.co.uk/Documents/IN%20Atmosphere.htm

This description—taken from the website—explains it perfectly.

> *The vacuum lamp lights up brightly whereas the gas-filled version glows only dimly, even though they are both of the same wattage. It's easy to prove that the filament is losing heat to the gas-filling because if the tops of the two bulbs are touched, the glass on the gas-filled one is very much hotter—despite the fact that its filament is visibly so much colder. This often confuses people, but the reason for the colder filament is that the gas is conducting heat away and transferring it to the bulb wall above.*
>
> *The effect of the gas convection currents is also superbly visualised, because only the upper portion of the filament lights up due to the fact that heat rises. If the lamp is inverted the hot gases move to the other end of the bulb, and the glowing part of the filament also moves. A shimmering effect may be observed due to turbulence currents as the gas flows around the stem.*[10]

What this tells us is that the conductive and convective affects of the gas causes the filament to cool and the surface of the bulb to warm, (reducing the brightness of the bulb) thus causing an equalising effect as it transfers heat from the wire to the glass bulb.

This same effect is happening on planet earth, as the atmosphere transfers heat away from the surface and transfers it around the globe and to the atmosphere above where it gets emitted away into space.

This cooling effect of gases can be demonstrated with the following example.

---

[10] Ibid.

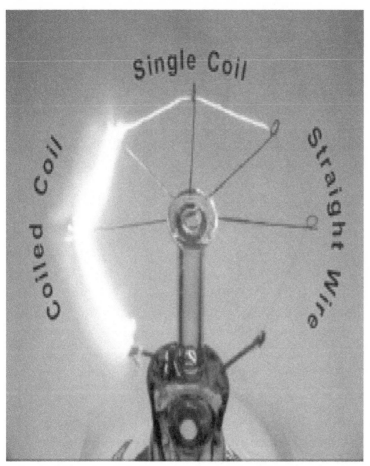

Diagram 3.4

Three in One Coils in a Gas[11]

In this picture, we see a bulb containing a gas. When the coils are arranged differently, the cooling effect of gases can be readily seen. For the straight wire, the highest rate of air cooling is experienced, but with a single coil the rate of cooling is less and so it glows brightly, but with a coiled coil, it glows the brightest.

---

[11] http://www.lamptech.co.uk/Documents/IN%20Coiling.htm

However, when this same test is done in a pure vacuum bulb, all three types of wire glow with equal brightness because they all reach the same temperature due to the current flowing through them. All three wires have only radiation to cool them, so they glow equally as bright because they don't experience different rates of convective cooling, because there is no convective cooling.

There is a detailed technical description on the lamp-tech website which explains why coiling a wire, causes less contact with the air and thus causes lower convective cooling if you are wondering why convective cooling differs between them.

It is an absolute certainty that the presence of the gas is having a convective cooling effect on the wire. This fact is known industry wide and manufacturers make use of this information to better design their bulbs to increase life expectancy.

So what we can then say is, if we had a vacuum light bulb with a straight wire as the lighting filament, it would glow brightly, but if we then added a carbon dioxide gas to the bulb to replace the vacuum, it would glow less brightly. It would do this, because the CO_2 is causing the wire to cool, because it is taking heat away from the wire due to conductive and convective effects.

Greenhouse gas frizzle frazzler's would have us believe in some back-radiation nonsense where the presence of the gas would actually cause the temperature of the filament to increase due to back radiation from the gas back to the filament, causing it to glow more brightly and to get warmer.

That is nonsense and something which simply does not happen. What happens is the gas warms and the coil cools as energy is transferred from the wire, into the gas which then gets transferred to the surface of the bulb.

Conclusion

1. Light bulbs emit short wave and long wave infra-red energy, which is directly absorbed by the CO_2 gas. (Not per the greenhouse theory)

2. The bottles should be sealed correctly, in both cases.

3. Carbon Dioxide is denser than air, as is Argon and Chlorine, yet no one does comparison experiments and shows they have higher temperatures too.
4. The bottles may or may not contain any water as this can skew the experiment.

5. Carbon Dioxide has a specific heat capacity less than air so it will be warmer if the same amount of energy is inputted. But then, so does Argon and Chlorine and these aren't greenhouse gases and yet they yield higher temperatures in these type of experiments.

6. The light bulbs (WHICH SHOULDN'T EVEN BE USED) should be positioned the same way for each bottle, but they are always seen skewed to favour the CO_2 bottle.

7. The temperature probes needs to be exactly in the centre of each bottle, they never are.

8. Manufacturers deliberately add gases to light bulbs—this can make them cooler and less bright, even if CO_2 is used in the bulb, the complete opposite of greenhouse effect theory.

There is no point in pretending that this two bottle experiment is scientific proof of greenhouse gas theory because it is not. It is in fact a direct refutation of it.

Professors, pretend climate scientists (actually religious zealots), tax hike mad socialists and wannabe communists kings, con-artists, goody two-shoe weak-wristed fakers all use this experiment to "TRICK!" people into believing that CO_2 causes devastating climate change and that the only cure is cost inducing systems of governance, control and unnecessary high taxes, which of course are used to pay their bills and further spread their lies.

Now you know the truth, you know the tricks used in the "Magic Show."

You are now in a position to demand educational authorities to remove the lies from the system and this is what you should do.

You are now in a position to defend yourselves from the liars in the political establishment who wish to trick you into raising taxes and introducing systems of governance which we don't need, so they can enslave populaces and control them for their own needs. And you should repel them and repeal all CO_2 and climate change rules, protocols and laws.

It really is all "A BIG LIE!"

It's time to wake to the real threat—from an ignorant populous mislead by overwhelming propaganda from the left.

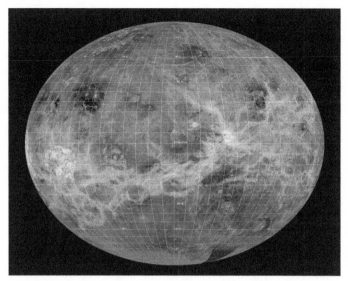

Illumination 4—What's Really Going on with Venus[12]

Executive Summary of Illumination

I WILL EXPLAIN that the following three factors are the real reason for Venus high temperatures.

- High atmospheric pressures, some 90 times greater than Earth.
- Highly volcanic and actively eruptive environment. "Thousands of active volcanoes."
- Thin crust, enabling transference of heat from the magma below to the atmosphere above.

Little Light Reaches the Surface of Venus

Greenhouse gas Frizzler's, would have you believe that a runaway greenhouse gas effect has occurred on Venus and this is why it's

---

[12] NASA, 2017

atmospheric and surface temperatures are so high. God, this is just silly beyond imagining.

Let's remind ourselves of how the "Green Fantasists" think a greenhouse warms. They wrongly and quite stupidly think they work by radiation where energy reaches the surface, is absorbed and converted into heat, with this heat being reflected back by the glass.

Well, let's look at this first point. Mustn't energy first reach the surface for it to then be reflected back? On Venus, this can't possibly be the case. All those whom have studied Venus quite rightly state that only a minimum of light from the Sun reaches the surface of that planet. Examples are provided below.

> *Venus albedo is about 0.60 (the amount reflected light in %), hence most of the light will never reach the surface.* [13]

Or...

> *About 1% of incident light reaches the surface.* [14]

Or...

> *The clouds we see on Venus are made up of sulfur dioxide and drops of sulfuric acid. They reflect about 75% of the sunlight that falls on them, and are completely opaque. It's these clouds that block our view to the surface of Venus. Beneath these clouds, only a fraction of sunlight reaches the surface.* [15]

Or...

> *Modern scientists can observe Venus' surface with radar, which*

---

[13] Nova Celestia, 2017

[14] *Fundamental Astronomy*, Hannu Karttunen, Fifth Edition, p 159

[15] Universe Today, 2017

penetrates the clouds.[16]

There is no disputing this knowledge. So, if little to no light reaches the surface, how can the warming of the atmosphere and the surface be as a result of "Reflected Heat Energy" if the energy isn't even reaching the surface? Well, quite obviously the answer to that question is that it can't. Frizzlers the world over will tell you that it most definitely is and pretend their "BIG LIE" is science, when it is not. It is high taxation preaching.

Far more likely that warming would occur as a result of "Direct absorption." Except, we already know that CO_2 doesn't absorb sunlight energy very well, so we also know that direct absorption by CO_2 can't be the cause, especially because the clouds are so reflective.

So, Why is Venus so Hot?

Venus Has High Atmospheric Pressures

The high density of the atmosphere allows the pressure at the air surface to reach 90 times that of the earth. The surface temperature of Venus never reaches below 400 degrees Celsius.[17]

It's well recognised that high atmospheric pressures, especially in summer, can cause warmer conditions here on Earth...

High pressure in the summer often brings fine, warm weather. It can lead to long warm sunny days and prolonged dry periods. In severe situations this can cause a drought.[18]

---

[16] Chicago Tribune, 1992
[17] Nova Celestia, 2017
[18] The Met Office, 2017

The combined gas law helps provide an explanation as to why high air pressures can result in higher temperatures.

Here's the ideal gas law equation:

$$\frac{P1 \cdot V1}{T1} = \frac{P2 \cdot V2}{T2}$$

What this tells us is if the volume of air remains equal and we increase the air pressure, then the temperature must rise.

So, for example, if we increased the air pressure from one standard atmosphere to two:

$$\frac{1 \cdot 1}{273} = \frac{2 \cdot 1}{T2}$$

T_2 must equal 546K for the equation to balance. This can be demonstrated simply by forcing more and more air into a container, even with the air pumped at room temperature. As we pump more and more into a fixed container, the air increases in temperature, simply because there are more molecules bouncing around the container, which means more collisions.

In the real world environment, the equation isn't as simple as this, because the air is free to expand and isn't trapped within a solid container, so the volume varies also. But, the basic principle still holds.

If you increase atmospheric pressure, then you can expect higher temperatures. Venus has high atmospheric pressure because there is just so many more molecules in its atmosphere when compared to Earth's atmosphere, probably as a result of the continuous volcanic eruptions and the lack of ocean to soak it all up.

With all those molecules constantly bumping into each other there is much more frictional heat generation and with all the extra molecules the air density and pressure is much higher than on Earth. This is what causes the high atmospheric temperatures.

However, we have many examples on Earth where higher air

pressure causes higher temperatures. For example, take Death Valley...

> *The biggest factor behind Death Valley's extreme heat is its elevation. Parts of it are below sea level.*[19]

Or, Turpan Depression, China...

> *This trough is the Earth's third lowest point reaching an elevation of 505 feet below sea level (-154 meters.) Located in China's western desert region south of Mongolia, the Turpan Depression is the country's hottest and driest area.*[20]

We can take the opposite approach to determine the truth also.

If we go higher in the Earth's atmosphere, it cools as the atmospheric pressure drops. At sea level, air pressure is one standard atmosphere or 1,013.25 Millibars.

As we rise through the atmosphere, the pressure drops because gravity pulls air downwards, so most air is at ground level. At 10,000 feet altitude the air pressure is roughly 700 Millibars and at 50,000 feet it is just over 100 Millibars.

What this means is that as we get higher, there are fewer and fewer molecules and therefore more space between the molecules. The fewer molecules there are, the less collisions there are between the molecules and therefore there is less pressure and also less energy and so therefore a lower temperature, as explained by the ideal gas law. Their kinetic energy is simply spread more thinly. Air molecules at ground level are crowded together like people huddled in a city and air molecules up in the tops of mountains are more spread out like in people in rural communities.

When was the last time you heard a "Green Fantasist" tell you that Venus was warm because of its air pressure? This is the last

---

[19] Live Science, 2017
[20] Wander Wisdom, 2017

thing they will tell you. Some will even claim that atmospheric pressure has no effect and that all the warming is purely down to reflected energy caused by the pretend greenhouse gases.

Reflected energy? What reflected energy? The sunlight barely reaches the surface.

I'd like those liars to stand inside a locked high atmospheric pressure chamber. I'll keep pumping the air in at room temperature and we'll see how long it takes them to bang the door asking to be released before it gets so hot it burns their skin.

There is no disputing this very basic fact. High atmospheric pressures make for high atmospheric temperatures. Venus has very high atmospheric pressures and this is a major reason why Venus is so hot.

It's nothing to do with emissive gases at all.

Don't let the "BIG LIE" confuse you.

Venus is Volcanic

This is self-explanatory. If you have thousands of volcanoes spewing heat and lava into the atmosphere, this will explain the high CO_2 levels, the high sulphur levels in the atmosphere and the high temperatures. The Venetian core is literally dumping heat into its atmosphere. Here are some examples.

> The surface of Venus is dominated by volcanism and has produced more volcanoes than any other planet in the solar system. The planet Venus has a surface that is 90% basalt, and about 80% of the planet consists of a mosaic of volcanic lava plains, indicating that volcanism played a major role in shaping its surface.[21]

> According to our modelling, the flank lava flows are the ones responsible for this [hot]spot. This is particularly important

---

[21] Crystal links, 2017

because this is the first time we can map, with such a high resolution, lava flows from a volcanic structure which is believed to be recently or still active on a terrestrial body other than Earth.[22]

There is no denying that Venus is highly volcanic as this extract helps demonstrate. "Volcanic activity is the dominant process for shaping the landscape of Venus, with over 90% of the planet's surface being covered by lava flows and shield volcanoes.[23]

There are many images of volcanoes on Venus which can be found on the internet.

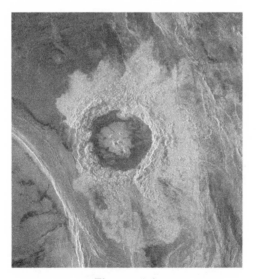

Figure 4.1

Dickenson Impact Crater and Lava Flows [24]

---

[22] Space.com, 2017
[23] Geology.com, 2017
[24] NASA, 1990

Frizzlers will no doubt be all up in arms. "How can you claim volcanoes causes Venus to be hot, you crazy denier?" You can see it, can't you. You can hear their whiny whinging ringing in your ears already. They are so desperate they will come up with anything to dispel the truth, which is that the "Greenhouse effect" is all "A BIG LIE!"

Seriously, is anyone really going to argue that volcanoes will make the surface cooler? The surface of Venus is littered with them—there are thousands of them.

> *There are huge lava flows hundreds of miles long and large plains created when lava filled low-lying areas. There are more than 100,000 small volcanoes and hundreds of large ones.*[25]

Just imagine standing near this. I don't think your first thoughts would be about donning a jumper to keep out the chill? Snow perhaps?

Figure 4.2

Imagery of Volcanism[26]

---

[25] Cornell University, 2017

[26] Pixabay, 2017

Vast swathes of Venus looks a lot like this. Stop giving "Frizzlers" credence.

Those Frizzlers will attempt to convince you that volcanoes erupted millions of years ago and are now dormant. That's also a lie. Here is the truth, as expressed by the people whom sent Venus Express.

If you see a sulphur dioxide increase in the upper atmosphere, you know that something has brought it up recently, because individual molecules are destroyed there by sunlight after just a couple of days.[27]

If sulphur dioxide is destroyed by just a few days sunlight, then there would be little to no sulphur dioxide in the Venetian atmosphere, however it is present in abundant quantities and so it must be getting there somehow. What would be the most probable cause? Why that would be active volcanoes. Such a theory was postulated back in the 80s.

Thus the amount of aerosols injected into the Venus middle atmosphere is greater by at least an order of magnitude than that associated with the most recent volcanic episodes on Earth.[28]

What other possible contributor for Sulphur Dioxide is there? When you think about the whole problem logically, it makes perfect sense. If the Sulphur Dioxide is destroyed within a few days, there must be active volcanoes replenishing its presence, otherwise there would be none.

Of course, "Frizzlers" have the problem of needing to toe the party line and say that "Global Warming" and "Climate Change" is a

---

[27] European Space Agency, 2012
[28] Larry W Espositio M, Copley, R Eckert, L Gates, A.I.F. Stewart & H Worden, 1988

man-made issue and that only scientists can understand it and deal with it, whilst of course, everyone else pays for it.

The next time you hear someone tell you Venus is hot because of pretend greenhouse gases, just politely remind them of the thousands of volcanoes on the planet, many of which are still active.

Oh, and if they start talking twaddle about a "Nuclear Winter" caused by the gas clouds blocking out the Sun making it cold, remind them that the atmosphere is 90 times the pressure of Earth and there's much more than just one active volcano on Venus. And if that wasn't enough evidence in itself, you should also know another very important fact…

Venus has a Thin Crust

Not many people know this, but compared to the Earth, Venus has very thin crust. The main reason is that it doesn't have an ocean to cool the lava flows from its magma core. If it did, the water would cool the lava flows rapidly and cause a thick layer to build over time. Also, oceans exert a huge weight on the crust, causing the plates to sub-duct down into Magma below, further cooling the core.

An atmosphere, even at high pressure, can't match this weight. But because they don't cool rapidly, a thick layer has no chance to build and so the heat and CO_2 just gets dumped into the atmosphere.

A thin crust means a direct heat transfer from the magma and core below as there is quite simply less insulation to the heat. This direct transfer of heat causes a massive warming to the Venetian atmosphere, which simply doesn't happen on Earth because the Earth's crust is thick.

The mean thickness of the crust is constrained to a range of 8-25 km, somewhat lower than previous estimates.[29]

---

[29] Peter B James, Maria T Zuber & Roger J Phillips, 2013, p859-875

Is anyone going to argue that having a thin crust—underneath which contains billions of tons of hot magma—isn't going to have warming effect to the atmosphere? The heat is of course going to transfer from below to above, just what else is it going to do?

Another paper from 1982 came to a similar conclusion: that the core of Venus must be losing its heat via some mechanism. And in an absence of water, its lithosphere (crust) must be thin...

...and forms resulting from plate convergence and divergence on Venus would differ substantially from those on the earth because of the high surface temperature and the absence of oceans on Venus, the lack of free or hydrated water in subducted material, the possibility that subduction would more commonly be accompanied by Lithospheric delamination, and the rapid spreading rates that would be required if plate recycling removes a significant fraction of the internal heat. [30]

This means Venus' core must be ridding itself of its heat somehow, or else it would just keep warming.

On Earth, tectonic plate movements and the massive body of oceans act as natural methods for removing this heat. We have seen the cooling effect of water on Lava flow, how it makes them solidify far more rapidly than on land. These oceans don't exist on Venus, only the extremely dense air is present to cool the lava flows, and because this air is so hot from its high pressure, little cooling to lava occurs, so therefore, other mechanisms must play more dominant roles. One mechanism maybe more dominant than the other or both might play equal roles.

But in any event, it is these mechanisms which are dumping heat into the Venetian atmosphere from the extremely super-hot magma and core below.

---

[30] Solomon, Sean C, Head, James W, 1982, p9236-9246

Plate Tectonics

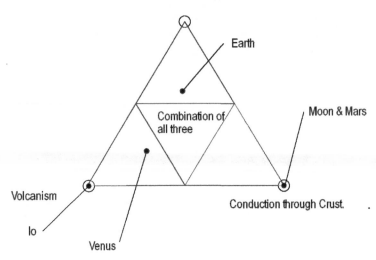

Diagram 4.3

Different Mechanisms for Planetary Core Heat Loss

This is a drawing I put together which briefly sums it up very well. All planets lose their heat by one of the three mechanisms, plate tectonics, conduction or volcanism. Here we can see that the Earth has a combination of all three effects cooling its core, but with plate tectonics playing a more significant role than the other two. We know we have many volcanoes both above ground and under the water, we have plate tectonics movement and we of course also have conduction but through our much thicker crust. The Moon and Mars are not known to have any active volcanoes on them and they don't appear to have tectonic plates. Io is a moon around Jupiter. This is how NASA describes it...

> *Jupiter's innermost moon, Io, is the most volcanically active body in our entire solar system! NASA missions imaged massive plumes shooting hundreds of kilometers above the surface, active lava flows and walls of fire associated with magma*

flowing from fissures.[31]

Venus, must cool via a combination of volcanism and lithospheric conduction, so I have shown it between the two.

Did a "greenhouse fantasist" ever tell you this when talking about Venus, or did they spout off about a "runaway greenhouse effect that will happen here on Earth, with the solution being, YOU PAY MORE TAX!"

Just plain odd, isn't it?

A bizarre and highly ineffective conclusion to a non-existent problem. There are no greenhouse gases, there is no greenhouse effect, there are no examples of runaway greenhouse gas planets. It's all nothing but "A BIG LIE!"

Conclusion

There is no point anyone ever using Venus as an example of greenhouse gas warming. Now you know the true reasons for Venus's high atmospheric and surface temperatures are its insanely high atmospheric pressure, thousands of volcanoes and its highly active eruptive environment coupled with the thinness of the Venetian crust.

Not a "runaway greenhouse effect."

---

[31] NASA, 2017

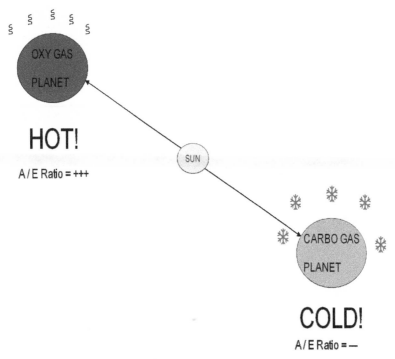

Illumination Five

Two Gas Planets Comparison

Executive Summary of Illumination

NOW I WILL show how it is "COMPLETELY ABSURD" to think that a Carbon Dioxide gas planet would be warmer than an Oxygen gas planet, especially at Earth orbit distance.

In my illumination two, I showed that adding a greenhouse to a plate, in a fully vacuum environment, caused the plate to cool and explained that this was because the area of emission had increased, whilst the area of absorption had remained the same. So this caused the energy to be spread more thinly, therefore resulting in lower temperatures and used an approximation calculation to show the lower expected iso-thermic temperature.

As you recall I explained that reducing emission areas would cause temperature of the plate to rise, and increasing emission areas would cause temperatures to fall.

Now, there are other ways of causing temperature changes in an object in space, whilst keeping areas exactly the same. I will explain brief examples below.

Example 1—Increasing Temperatures by Reducing IR Emissivity

Using the two-sided flat plate example again, we can cause the temperature of the plate to rise by selecting a different material, one with a lower emissivity.

A standard flat plate, with ε (IR emissivity) = 1.0 would maintain steady state temperatures of 331K or 58 C°.

When the IR emissivity is 1.0 and the solar absorptivity is 1.0, this is known in Rocket Science parlance as an A / E ratio of 1.0.

Now let's suppose we selected an imaginary material with a lower emissivity 0.5 but maintained solar absorptivity of 1.0. The plate would have a steady state temperature of 394K or 121 C°. This is because the material doesn't emit energy so readily and so needs to achieve higher temperatures to emit the energy that it absorbs. I.e. If the plate is absorbing 1367 watts of energy each second, it will emit energy out at 1367 watts at 121 C°. This is because at 331K the plate is only emitting a total of 683 watts, so its keeps on warming until it achieves equilibrium with the amount of energy it is absorbing.

To turn this example to gases, if a gas planet was less emissive, compared to a gas planet which was more emissive, the less emissive gas planet would also need to rise in temperature to emit out the energy it absorbed.

This object has an A / E Ratio of 2.0. Objects with A / E Ratios higher than 1.0 maintain steady state temperatures which are higher.

A point to note, is that Oxygen and Nitrogen are both far, far less emissive than CO_2 which is highly emissive. So in the absence of any other inputs, if we had an oxygen gas planet and a carbon dioxide gas planet (with no rocks at the centre) and if we assumed they both absorbed solar energy equally, we can understand that the Oxygen and Nitrogen planets would need to be hotter than the CO_2 planet, to emit out the same heat, because their molecules don't emit IR radiation, so as they absorb energy the temperatures keep rising until they reach temperatures at which they do emit radiation so that the rate of energy being emitted matches the rate of energy that they are absorbing.

Example 2—Decreasing Temperatures by Reducing Solar Absorptivity

Going back to the two-sided flat plate, if we assumed that it's IR was 1.0 and its solar absorptivity is one, we know we have a steady state temperature of 331K.

If we now select a material which has a lower solar absorptivity, $\acute{a} = 0.5$, what happens is this material isn't very good at absorbing solar rays, the rest just reflect off. This has the effect of reducing the energy input.

So sticking with the two sided plate, the effect is the steady state temperature reduces, to 279K or 6 C°. As only 684 watts is being absorbed, and the plate emits 342 watts from each side. Easy to follow, isn't it.

This object has an A / E ratio of 0.50. Objects in space which have A / E ratios lower than one, will maintain steady state temperatures which are lower. So an object in space with a high A / E ratio is going to be hot and one with a low A / E ratio is going to be cold. You can see the comparison in Diagram 5.1

To turn this example, to gases if we had a gas which had low solar absorptivity and compared that to a gas with high absorptivity, we know that the gas planet with the lower absorption ratio, is going to be cooler.

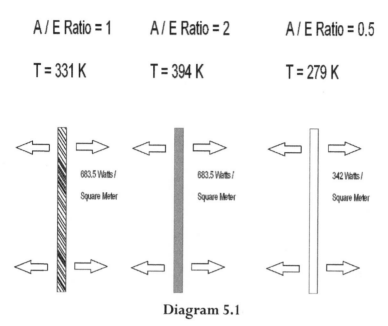

Diagram 5.1

Comparison of three plates with different A / E Ratios

Comparing, Oxygen with CO_2, we know that Oxygen absorbs Ultra-Violet, it also absorbs some green light and red light spectra. CO_2 on the other hand doesn't, it has a very low absorption of Sunlight. This means that Oxygen has a higher absorption than CO_2, meaning it will maintain higher temperatures.

But here's the big clincher, CO_2 does emit equally with Oxygen. CO_2 is far more IR emissivity than Oxygen and CO_2 is simultaneously a poor solar absorber and Oxygen and relatively good absorber. How do we resolve this?

Example 3—The Importance of A / E Ratios

Going back to the two-sided flat plate. If we pretended that the flat plate had the same low absorption of CO_2 and high emissivity of a planets worth of CO_2 what we would have would be something along the lines of $\acute{\alpha} = 0.003$ and $\acute{\epsilon} = 0.15$.

So the flat plate would have a seriously low A / E Ratio of 0.02. This means it is going to be seriously cold in space. The plate would maintain a temperature of just 125K or -148 C°.

If we now pretended that the flat plate had similar properties to a planets worth of Oxygen, what we would have is something along the lines of $\acute{\alpha} = 0.13$ and $\acute{\epsilon} = 0.003$ A / E ratio of 43.3, one seriously hot plate. This plate would maintain a temperature of 850 K or 557 C°.

Here is a table which gives a range of materials with different A / E ratios and in the notes column you can see it makes reference to the fact that aluminium foil would be very hot, because it has high A / E ratio. This isn't made-up science, this is the real science at hand which needs to be considered.

So you can see that placing a rock inside a freezing cold gas planet is not going to cause the rock to warm, quite the reverse. Therefore, "Greenhouse Gas Effect" theory can only be completely flawed. It is a total nonsense.

	aS	E	
	Solar	Surface	aS/E
Material	Absorption	Emissivity	ratio
Silver, Highly polished		0.02 – 0.03	3.00
Gold, Highly polished		0.02 – 0.04	3.00
Barium Sulphate with Polyvinyl Alcohol	0.06	0.88	.07
Aluminum polished	0.09	0.03	3.00
Magnesium Oxide Paint	0.09	0.90	.10
Magnesium / Aluminium Oxide Paint	0.09	0.92	.10

Aluminum quarts overcoated	0.11	0.37	.30
Aluminum, Highly polished		0.04 – 0.06	3.00
Snow, Fine particles fresh	0.13	0.82	.16
Zinc Orthotitanate with Potassium Silicate	0.13	0.92	.14
Aluminum anodized	0.14	0.84	.17
Aluminum foil[32]	0.15	0.05	3.00

Table 5.1 of A & E Ratios[33]

Example 4

Oxy and Gas Planets

Planets aren't flat, they are spherical, this means adjustments need to be made to the absorption and emission areas.

Without showing all the maths, you would expect "Carbo" have an average temperature of T = 105 K or -168 C° Where as an "Oxy" gas planet would have an average of 715 K or 442 C°. Anyone who thinks differently is free to postulate what they think the average temperatures of these two imaginary gas planets would be (just gas, no rock.) You will come to the same conclusion, "Oxy" hot, "Carbo" cold.

---

[32] Aluminum foil gets very hot because of this high ratio.
[33] Solar Mirror, 2017

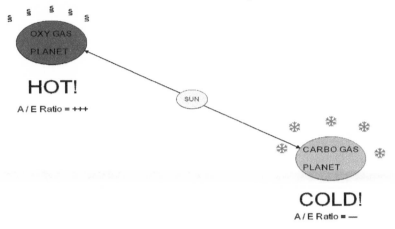

Diagram 5.2

The Two Gas Planets

Obviously in a real world environment we would expect much higher temperatures high in the atmosphere and much lower temperatures lower down due to the initial absorption of radiation at the surface layer and cooler closer to the centre if all the air was magically equally spread and didn't mix around. Also, as explained previously, the gravity of the rock would pull the air down to the centre which would increase the air pressure and hence would increase temperatures as well. So we would see, high temperatures near the surface, with a cooling as altitude in the atmosphere rises followed by a warming again high up in the atmosphere as direct absorption of solar radiation causes very high temperatures. Similar to the Earth's atmosphere.

So if we then got a perfectly spherical rock, with an A / E ratio of 1 with an average temperature of 279K and stuck it the hot "Oxy", we can see how the rock would warm. Conversely, we can see how the rock would not only cool, but completely freeze if we stuck it in the middle of the "Carbo" gas planet. And those thick wit, CO_2 causes global warming "Frizzle Faces", expect you to believe the complete and utter reverse. They think placing a rock in a freezing gas would cause it to warm! Like I said before.

"COMPLETELY ABSURD!"

Conclusion

Now you also know that a pure Oxygen planet would be warm and why that would be and a pure CO_2 planet would be freezing cold and why that would be. And to place a planet sized rock at the centre of "Oxy", would cause the planet to warm and yet placing a planet sized rock into "Carbo" would cause the planet to cool. To think anything else is just plain stupid, like I said in my last letter, it is "COMPLETELY ABSURD".

So replacing Oxygen in our atmosphere with CO_2 just can't cause a warming, CO_2 is an emissive gas and thus cause's localised atmospheric cooling, nothing else. Therefore if there really is any warming occurring on the planet, it is not "because of" increased levels of CO_2 but "in spite of".

To think planetary temperatures can be controlled by playing around with CO_2 levels, is ridiculous, therefore resources should not be wasted in such fruitless and pointless pursuits and instead money be spent in other more effective ways.

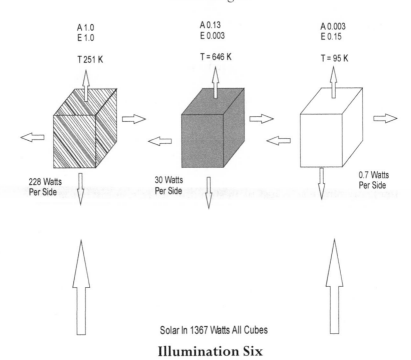

Solar In 1367 Watts All Cubes

Illumination Six

De-frazzling the Frizzler's Frazzle using Cubes and Cuboids

Executive Summary

WHEN A CLIMATE alarmist, the Frizzler's, try to bedazzle you with blinding science, telling you that harmful back radiation rays are going to frazzle the Earth, leaving everything asunder and barren you need to be able to provide examples which show how utterly ridiculous they are. And these are examples in that vein on how to do that. You should be aware that this illumination is a bit more brain busting than the previous ones, if you have read the previous ones you should still be able to cope.

In this Illumination I will show:

- The temperature of three stand-alone cubes in space.
- Imagine touching these cubes.
- I form rectangular cuboids and show the temperatures.
- I show how emissivity is the dominant factor in lowering temperatures, not increasing temperatures.
- I show the view factor effects of separated cubes using view factor maths.

The Temperatures of Three Stand Alone Cubes

Imagine three stand-alone cubes in space, no-where near each other and so un-able to affect each other all at Earth orbit distance from the Sun. One a perfect black body A / E ratio 1 (hatched cube), another with a very high A / E ratio 43.3 (the grey cube) and then another with a very low A / E ratio 0.02 (the clear cube).

The temperature data is all shown in the Table 6.1 and on Diagram 6.1.

What we see is the black body cube maintains steady state temperatures of 252K, the high A / E Cube maintains temperatures of 646 K and the low A / E cube maintains temperatures of just 95K.

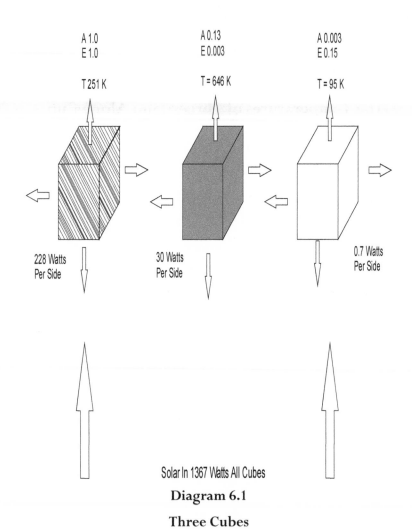

A 1.0
E 1.0

T 251 K

A 0.13
E 0.003

T = 646 K

A 0.003
E 0.15

T = 95 K

228 Watts
Per Side

30 Watts
Per Side

0.7 Watts
Per Side

Solar In 1367 Watts All Cubes

Diagram 6.1

Three Cubes

ITEM	Cube 1- Blackbody	Cube 2 – High A/E	Cube 3 – Low A/E
Absorptivity	1	0.13	0.003
Emissivity	1	0.003	0.15
A / E Ratio	1	43.3	0.02
A(e)	6	6	6
Temperature (K)	251	646	95
Q Per Side (Watts)	228	30	0.7
Total Q (Watts)	1367	178	4

Table 6.1

Temperature Calculations for Three Separate Cubes

If you were flying around space and you happened to come across these three cubes, you would get very different results if you decided to touch them.

Cube 1 has the highest rate of absorption, absorbing all the energy and the highest rate of emission, as it emits it all back out again yet it maintains a temperature of only 252K. The Grey cube however maintains the highest temperature, yet absorbs much less energy and so also emits out much less energy. If we looked at this cube through an infra-red device, it would look cold because of its low emissivity, despite being very hot. This phenomenon is known as "APPARENT TEMPERATURE" and this can cause a lot of confusion to people whom are unaware of it. Imagine touching this cube, it would burn you. Apparent temperature is when something, appears to be one temperature via an IR camera, but it actually is another, usually much higher temperature.

A lot of people confuse energy, temperature and heat as all being the same thing they are not. As you can see with these easy to follow clear cut examples, the hot Grey cube, emits less energy than the black body but it is much hotter. Don't let Frizzlers, "Frazzle" your brain with non-sense. Remember this, bodies do not contain heat, but think of heat as something that flows from system boundaries. Heat is how we change the energy of a system, because of the temperature differences produced by the system. I.e. Hot body to cold body, allows heat transfer from the hot to cold, equalizing the temperature in both bodies.

The third and final cube, the clear cube, not surprisingly is the coldest at a mere 94.68 Kelvin or -178.32 Celsius. It has the lowest solar absorption powers and so has little energy to excite the atoms of the cube. It also has a much higher level of emissivity compared to the Grey cube and so the little energy that it does absorb is thrown away. This means that it is very cold. If you touched this, the rate of heat transmission from your warm hand would be so fast that you would risk freezing it altogether. In space, without some sort of protective gear, this is something you would not want to handle.

Imagine having a business partner, the Grey cube would represent a partner that was very good with money. Most of the money you would give it, it would save and it would keep saving until it had a big balance, before spending the money, whereas the Clear cube would represent a partner that was very bad with money and spent everything almost as quickly as they got it and so would result in a very low bank balance.

In these examples, the hot cube is a symbol of representation for a non-emissive gas and the cold cube is a symbol of representation for an emissive gas. We can clearly see that when something is more emissive, it results in colder temperatures. This is the reason why pipework insulation is usually wrapped in aluminium, because its low emissivity helps keep in the heat. Has building engineering the whole world over really been getting it so wrong for all this time? I think not.

Rectangular Cuboid with Both Boxes Absorbing Sunlight

Now it's time to start making things more interesting.

In this comparison I am going to compare, three cuboids all of the same dimensions. Cuboid 1 is a pure-blackbody cuboid, Cuboid 2 is a combination of a black cube and a Hatched (Less emissive) cube and Cuboid 3 is a combination of a black cube and a Clear cube (more emissive).

What effect would this have on the expected temperatures? In order to make these Cuboids Isothermic, I have assumed a very high conductance, of 2000 w/mk. You could also assume a full vacuum, this would have the same effect. This high rate of conductance will mean large quantities of heat, at a rate even greater than radiation can transmit through the material.

Starting with the blackbody cuboid, we can see that T = 263.51K. This is because it has a higher ratio of absorption area compared to its emission area, than just a stand-alone cube. So for a cube, the absorption area was $1m^2$ and the emission area was $6m^2$, this gives a ratio of 1:6 or 0.167. Now that the absorption area has increased to $2m^2$ and the emission area has increased to $10m^2$ the ratio is now 1:5 or 0.20. What this means is as a proportion of the surface area of the body, more is devoted to absorption than before, this is why higher temperatures can be maintained.

You can see with this cuboid the solar energy in is 2734 Watts with each half emitting exactly 50% of the energy. This is exactly what you would expect for an isothermal blackbody cuboid of the dimensions shown in Diagram 6.3.1.

Now with the hatched combo cuboid, you can see that the actual amount of energy being absorbed is now less than blackbody cuboid. However, because half of this cuboid is made of a low emissivity material, the equilibrium temperature is higher than the blackbody. Here T = 271.48K or thereabouts.

Again, as in the first example, the Clear combo cuboid, because half of this cuboid has less absorption powers but stronger emission powers, it maintains the lowest steady state temperatures at approximately T = 254.65K.

You should start to see, that no matter how we arrange things, the presence of the low A / E ratio body, exerts a cooling influence and that increasing emissivity causes lower temperatures, not higher ones as "Frizzlers" would have us believe.

A point to note, if you look at Diagram 6.3.2, if I separate the cuboid by a distance of 1mm, the temperature of the two single cubes stays the same at 263.51. There is no back-radiation warming occurring because the temperature difference is 0, there is no increase in temperature. No "Frazzling" occurs, despite what "Frizzlers" would have us believe. Combining the two stand-alone cubes caused an increase in temperature from 252K each, to 263.51K for the cuboid as a result of increasing the absorption to emissive surface area ratios, enabling a greater amount of energy to be stored within the object.

	Clear Cube	Hatched Cube	Black-Body Grey Cuboid
α (c2)	0.003	0.15	1
e (c2)	0.15	0.003	1
T1	254.65	271.48	263.51
Q1 Solar	1,367.00	1,367.00	1,367.00
Q2 Solar	4.1	177.71	1,367.00
Total In	1,371.10	1,544.71	2,734.00
Q1 Out	1,192.22	1,540.04	1,367.00
T2	254.67	272.16	263.51
Q2 Out	178.88	4.67	1,367.00
Total Out	1,371.10	1,544.71	2,734.00
Proportions Out - C1	86.95%	99.70%	50.00%
Proportions Out - C2	13.05%	0.30%	50.00%

Table 6.3.1

Steady State Temperature Achieved with Both Sides of the Cuboid Receiving Light

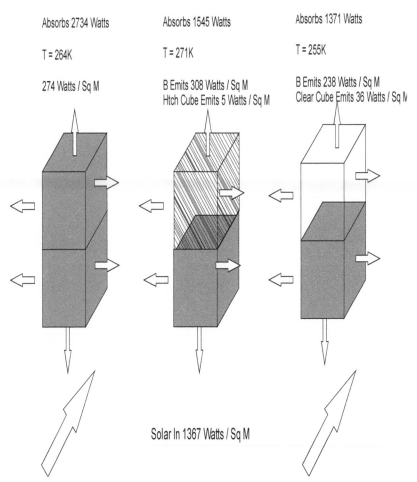

Absorbs 2734 Watts

T = 264K

274 Watts / Sq M

Absorbs 1545 Watts

T = 271K

B Emits 308 Watts / Sq M
Htch Cube Emits 5 Watts / Sq M

Absorbs 1371 Watts

T = 255K

B Emits 238 Watts / Sq M
Clear Cube Emits 36 Watts / Sq M

Solar In 1367 Watts / Sq M

Diagram 6.3.1

Cuboids Receiving Light Comparison

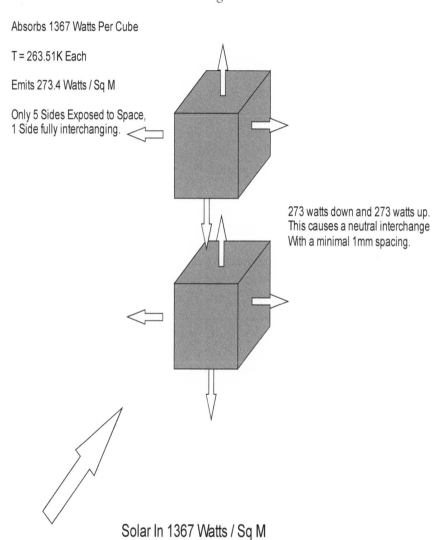

Absorbs 1367 Watts Per Cube

T = 263.51K Each

Emits 273.4 Watts / Sq M

Only 5 Sides Exposed to Space,
1 Side fully interchanging.

273 watts down and 273 watts up.
This causes a neutral interchange
With a minimal 1mm spacing.

Solar In 1367 Watts / Sq M

Diagram 6.3.2

**Blackbody Cuboid Split—No Radiation Warming
Experienced**

Rectangular Cuboids with One-Half in the Shade

In this comparison, we can see that the blackbody maintains equilibrium temperature T= 221.58 K, the hatched combo cuboid T = 263.32K and the clear combo cuboid T = 254.47 K. Here we see the relationship holds again, the object with the highest emissivity, maintained the lowest temperatures and the object with the lowest emissivity maintained the highest temperatures.

The blue combo cuboid, is warmer than the black body, because as an object it has a higher A / E ratio than the pure black body.

If you look at Table 6.4.1 and Diagram 6.4.1, you can see what I have done is combined the effects of two cubes, with each cube being capable of radiating from 5 sides, with the internal side not being able to radiate to space. The first half of the cube is the same in all circumstances, the second back cube I have shown the range of temperatures for differing emissivity's for this second cube.

At an emissivity of zero, if such a material were to exist, it would never lose any heat by radiation. Therefore all heat lost from the cuboid can only be lost from the first half of the cube from its emitting 5 sides. This effectively acts as a brilliant insulator to the cube and is nothing but a store of heat.

As you can see, if I start to increase the emissivity of the second cube, temperatures reduce as its insulating properties reduce, as this part of the cuboid is now losing heat. This extra heat loss, means that steady state temperatures lower as emissivity increases. You can see for example that at an emissivity of 0.5 from the second cube, it emits from 5 sides into space a total of 455.64 watts of heat and is responsible for a third of the heat emissions of the cuboid.

This is clearly the complete reverse of what "Frizzlers" would have us believe, that increasing emissivity of the atmosphere results in warmer temperatures of the planet.

The act of rotating the body and reducing the amount of sunlight received by the object can cause vastly different temperatures on the object.

The only reason the clear combo object is warmer than the blackbody, is because it has a comparatively lower emissivity than the full black body.

If I started to make the cube act more like an atmosphere, by introducing transparency of infra-red radiation to the second half of the cube I get interesting results as shown on Table 6.4.2.

In this table, I have made transparency to IR radiation, inverse to the cubes emissivity.

So for example an emissivity of 0, would be full transparency and an emissivity of 1, would be full absorption of radiation emitted from the 1st half of the cuboid.

Temp Diff	e	T1 (k)	Q1 (out)	T2 (k)	Q2 (Out)	Total Out	Proportions	
0	0	264	1367	264	0.00	1,367	100%	0%
0.19 (Hatch)	0	263	1363	261	3.93	1,367	100%	0%
4.19	0.1	257	1243	257	124	1,367	91%	9%
9.04 (Clear)	0.2	254	1189	254	178	1,367	87%	13%
11.74	0.2	252	1139	252	228	1,367	83%	17%
16.73	0.3	247	1052	247	315	1,367	77%	23%
21.26	0.4	242	976	242	391	1,367	71%	29%
25.4	0.5	238	911	238	456	1,367	67%	33%
29.21	0.6	234	854	234	513	1,367	63%	37%
32.73	0.7	231	804	231	563	1,367	59%	41%
36.01	0.8	228	759	228	608	1,367	56%	44%
39.06	0.9	224	720	224	647	1,367	53%	47%
41.93 (Grey)	1	222	684	222	684	1,367	50%	50%

Table 6.4.1

Cuboid Comparison, Half-Shaded

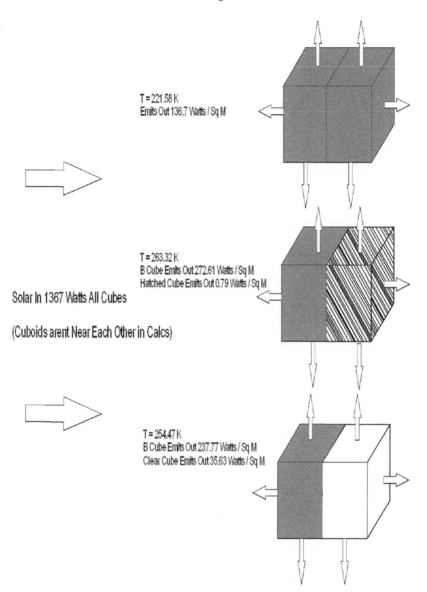

T = 221.58 K
Emits Out 136.7 Watts / Sq M

T = 263.32 K
B Cube Emits Out 272.61 Watts / Sq M
Hatched Cube Emits Out 0.79 Watts / Sq M

Solar In 1367 Watts All Cubes

(Cuboids arent Near Each Other in Calcs)

T = 254.47 K
B Cube Emits Out 237.77 Watts / Sq M
Clear Cube Emits Out 35.63 Watts / Sq M

Diagram 6.4.1

Combination Cuboid Comparison, Half-Shaded

Trans-parency	e	T (k)	Q1 (out)		T2	Q2 (out)	Total Out
			5 Sides	Thru 2			
1.0	0.0	252	1139	228	252	0	1367
0.997 (Hatch)	0.0	252	1137	227	249	3	1367
0.9	0.1	248	1068	192	248	107	1367
0.85 (Clear)	0.2	246	1036	176	246	155	1367
0.8	0.2	244	1005	161	244	201	1367
0.7	0.3	241	949	133	241	285	1367
0.6	0.4	237	899	108	237	360	1367
0.5	0.5	234	854	85	234	427	1367
0.4	0.6	232	814	65	231	488	1367
0.3	0.7	229	777	47	229	544	1367
0.2	0.8	226	743	30	226	594	1367
0.1	0.9	224	712	14	224	641	1367
0.0	1.0	222	684	0	222	684	1367

Table 6.4.2

Cuboid Comparison Half-shaded, with Full Transparency of the Rear Cube

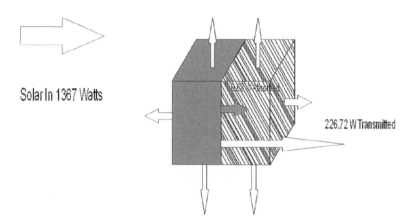

Solar In 1367 Watts

226.72 W Transmitted

T = 251.70 K
B Cube Emits Out 227.4 x 5 Sides Watts / Sq M
B Cube Transmits 226.72 through Hatched Cube
Hatched Cube Absorbs 0.68W from B Cube
Hatched Cube Emits Out 0.65 Watts / Sq M

Diagram 6.4.2 Cuboid Comparison, Half-Shaded, with Full Transparency of the Rear Cube (WARMER)

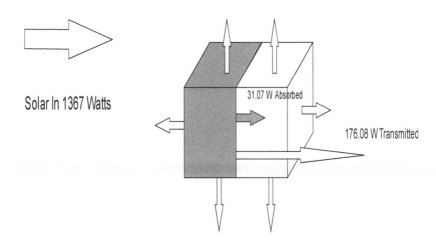

Solar In 1367 Watts

31.07 W Absorbed

176.08 W Transmitted

T = 245.90 K
B Cube Emits Out 207.15 x 5 Sides Watts / Sq M
B Cube Transmits 176.08 through Clear Cube
Clear Cube Absorbs 031.07W from B Cube
Clear Cube Emits Out 31.03 Watts / Sq M

Diagram 6.4.3 Cuboid Comparison, Half-Shaded, with Full Transparency of Rear Cube(COLDER)

Allowing for transparency, means that the 1st half of the cube can now lose IR radiation from all 6 sides. This is why we can see that the maximum temperature, which occurs at an emissivity of zero for the second cube is 11.74 K less than the non-transparent cuboid. This is because the first half, reaches steady state conditions whilst losing heat from 6 sides instead of 5. Again, as the second half of the cube has no emissions, it acts as a perfect insulator and is nothing but a store of heat, allowing only IR radiation through it.

Again, we see that the Hatched-Combo cuboid is warmer than the clear-Combo cuboid, despite the clear-Combo having the ability to absorb the emitted radiation. The amount of radiation lost through it is clearly less at 176.08 Watts, with hatched combo losing 226.72 through it. But, because this is gain of radiation energy, is more than offset by the fact that the second half of the cuboid emits out more radiation to space. The difference is made up of that which is transferred to it by conduction.

In the clear Combo example, the first cube emits a total of 1242.92 Watts at (e=0.15) with 5 sides emitting a total of 1035.77 watts into space and the 6th side emitting 207.15 Watts a total of 1,242.92. Of the 207.15 only 31.07 Watts is absorbed by the second cube and the remaining 176.08 is lost out into space. Steady state conditions therefore occur at roughly T = 245.85 whereby the second half cube with an (e=0.15) emits out a total of 155.15 Watts from all 5 sides exposed to space. The cuboid exactly emits 1367 watts with one half of the cube being able to emit at 6 sides with blackbody emissivity and the second half only being able to emit at 0.15 emissivity.

We see that, as with the solid cuboid, the temperature reduces as we increase the emissivity of the second half of the cuboid. The cuboid reaches a minimum temperature which exactly matches the non-transparent cuboid of 221.58, if the second half absorbs all outgoing radiation from the sixth side of the first cube.

What happens if I also factor in back-radiation from the second half of the cuboid, to the first half of the cuboid to steady state conditions? Absolutely nothing. This is because the rate of emission from the cuboid, must match the rate of heat input into the cuboid. If one side of the cuboid gets warmer, the other side must get cooler for steady state conditions to exist and if one side is warmer than the other, then heat transfer occurs until temperature equilibrium is achieved. What does happen however, if I introduce thermal capacity into the cuboid, is that the first cube warms up more quickly, reaching its steady state temperature fractionally earlier than the second half of the cuboid and that's about it.

So in these examples, we can clearly see, increasing the emissivity of the second half of the cuboid, despite being in receipt of ever greater rates of surface emitted infra-red radiation, results in lower temperatures.

The exact same thing is happening in the atmosphere. Any vapours and gases which get put into the atmosphere, increase the emissivity of the atmosphere and therefore will result in cooler temperatures. However, CO_2 is present in such small quantities that its effect is negligible. Chasing rainbows trying to stop global warming, if such a thing is even happening, by restricting CO_2 emissions, is a project doomed to fail. Why participate?

Split Cuboids into Front and Back Cubes Separated by 1mm

Now let's see what happens if we split the cubes and maintain a distance of the back cube from the front cube of 1mm in order to maximise radiation view factor forcing effect on the first cube. What conditions will we arrive at during steady state conditions?

I have utilised the following formulae to determine the view factor forcing effect between the two flat sides each cube has exposed to each other and assumed they will of course be perfectly parallel and then performed a series of iterative calculations to find the eventual steady state temperatures.

$$F = (1 \ / \ \pi \, w^2) \ \{\ln \, [x^4 / (1 + 2w^2)] + 4wy\} \ ^{34}$$

Whereby:

$w = W \ / \ H$ (and $W = 1 \times 1$)
$x = \sqrt{(1 + w^2)}$
$y = x \arctan(w/x) - \arctan w$

To summarise the forcing effect I have shown the effect of distance on Forcing below for a 1m by 1m facing separated by distance h.

---

[34] Radiant Heat Transfer, 2008

Reduction of Forcing with Distance	
F	Distance
1.00	0.001m
0.98	0.01m
0.83	0.1m
0.20	1.0m
0.07	2.0m
0.00	10.0m

Table 6.5.1—Forcing Effect against Distance H

Here we can see that for a 1m by 1m cube full radiation forcing is achieved at 1mm which after a separation of just 10 metres, negligible forcing is achieved.

2 Cubes back to back – Black Bodies		
Absorptivity	1	1
Emissivity	1	1
A / E Ratio	1	1
Total Surface Area (m2)	6	6
Total no Sides	6	6
Separation (H) (m)	0.001	0.001
Area for solar absorption	1	0
Area for emission	6	6
Solar Energy In (W)	1367	0
Interchanged Energy (W)	234.34	39.06
Steady State Temp (K)	253.55	162
Energy Out 5 Side to Space	1171.71	195.29
Energy Through Other Cube	0	0
Total Out to Space (W)	1171.71	195.29
1367		

Table 6.5.2a—Steady State Temperatures of Separated Cubes—Black Bodies

Black Dragon

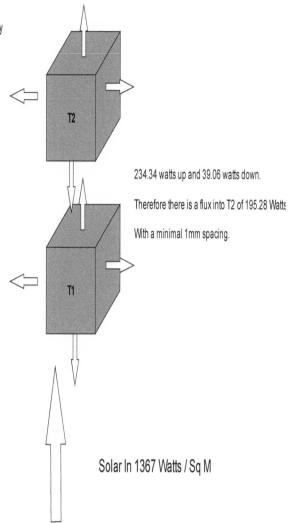

Absorbs 1367 Watts First Black Cube Only

Black T1 = 253.34 K

Black T2 = 162.00 K

Black 1 Emits 253.55 Watts / Sq M
Black 2 Emits 39.06 Watts / Sq M

Only 5 Sides Exposed to Space,
1 Side fully interchanging.

T2

234.34 watts up and 39.06 watts down.

Therefore there is a flux into T2 of 195.28 Watts

With a minimal 1mm spacing.

T1

Solar In 1367 Watts / Sq M

Diagram 6.5.2a

**Steady State Temperatures of 3 Combinations of
Separated Cubes—Black Bodies**

2 Cubes Back to Back—Hatched Combo		
Absorptivity	1	0.13
Emissivity	1	0.003
A / E Ratio	1	43.33
Total Surface Area (m2)	6	6
Total no Sides	6	6
Separation (H) (m)	0.001	0.001
Area for solar absorption	1	0
Area for emission	6	6
Solar Energy In	1367	0
Interchanged Energy	0.11	0.68
Steady State Temp (K)	251.77	160.87
Energy Out 5 Sides to Space	1,139.26	0.57
Energy Through Other Cube	227.17	0
Total Out to Space	1,366.43	0.57
1,367.00		

Table 6.5.2b

Steady State Temperatures of Separated Cubes—Black Body and Hatched Body

Black Dragon

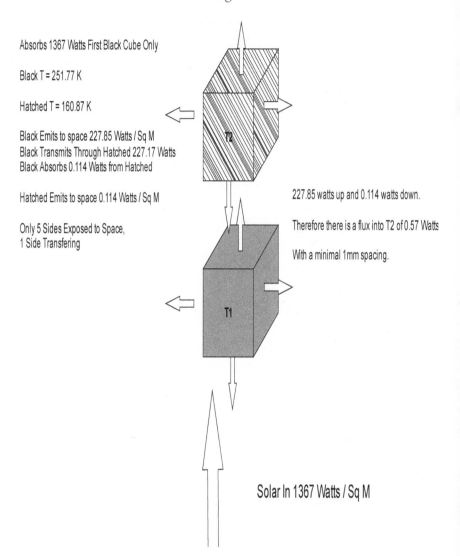

Absorbs 1367 Watts First Black Cube Only

Black T = 251.77 K

Hatched T = 160.87 K

Black Emits to space 227.85 Watts / Sq M
Black Transmits Through Hatched 227.17 Watts
Black Absorbs 0.114 Watts from Hatched

Hatched Emits to space 0.114 Watts / Sq M

Only 5 Sides Exposed to Space,
1 Side Transfering

T2

T1

227.85 watts up and 0.114 watts down.

Therefore there is a flux into T2 of 0.57 Watts

With a minimal 1mm spacing.

Solar In 1367 Watts / Sq M

Diagram 6.5.2b

**Steady State Temperatures of Separated Cubes—Hatched
Combo**

2 Cubes Back-to-Back—Clear Combo		
Absorptivity	1	0.003
Emissivity	1	0.15
A / E Ratio	1	0.02
Total Surface Area (m2)	6	6
Total no Sides	6	6
Area for solar absorption	1	0
Area for emission	6	6
Separation (H) (m)	0.001	0.001
Solar Energy In (W)	1367	0
Interchanged Energy (W)	5.72	34.32
Steady State Temp (K)	252.03	161.03
Energy Out 5 Sides to Space	1,143.93	28.6
Energy Through Other Cube	194.47	0
Total Out to Space (W)	1,338.41	28.6
1,367.00		

Table 6.5.2c

Steady State Temperatures of Separated Cubes—Black Body and Hatched Body

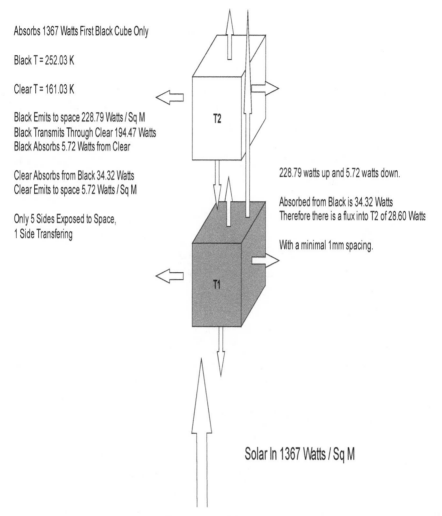

Absorbs 1367 Watts First Black Cube Only

Black T = 252.03 K

Clear T = 161.03 K

Black Emits to space 228.79 Watts / Sq M
Black Transmits Through Clear 194.47 Watts
Black Absorbs 5.72 Watts from Clear

Clear Absorbs from Black 34.32 Watts
Clear Emits to space 5.72 Watts / Sq M

Only 5 Sides Exposed to Space,
1 Side Transfering

T2

228.79 watts up and 5.72 watts down.

Absorbed from Black is 34.32 Watts
Therefore there is a flux into T2 of 28.60 Watts

With a minimal 1mm spacing.

T1

Solar In 1367 Watts / Sq M

Diagram 6.5.2c

**Steady State Temperatures of Separated Cubes—Clear
Combo**

Here in these split examples we have what the maximum view
factor radiation forcing could ever possibly be, using the equation
shown above, so I could determine what the affect is for this

arrangement. This forcing factor is then added into the equation to produce $Q = \varepsilon A(e) F \sigma T^4$. I then calculated steady state conditions as shown in Table 6.5.2_.

What we find is the 2 black body cubes give maximum surface temperatures and the Hatched-combo giving minimum temperatures and the Clear combo giving temperatures in between. The Frizzlers, will all be joyous, we have just proved the Frizzle Frazzle and showed it gives higher temperatures. We're all doooooooomed!!! Any minute now pictures of deserts and floods will start appearing along with Frazzle Factor Five proclamations of planetary destruction and devastation by Chief Frizzlers worldwide.

No doubt Al-Gore will start using this article as proof positive of the "Frazzle Force."

What small minds the Frizzlers have. So desperate to bring in their one world climate tax order, they will make their entire philosophy completely dependent on this tiny, negligible and completely irrelevant force.

Firstly these view factor calculation arrangements only work in a "full vacuum", when the objects are not in contact with each other. If there is contact between the objects, as occurs within the atmospheres of planets and their surfaces, then conduction, convection and latent heat transfers all radically change the picture, to one nothing like the separated cubes to one exactly like the Cuboid. Just think of the examples we had in <u>Diagram 6.4.1 & 2</u> This is because conductance, convection and latent heat transfer are able to transmit far more energy per square meter of substance than radiation can and so for example if radiation forcing was 4 watts, but heat input from conduction was 10 watts then there is never a net flow due to radiation forcing, so the effects are of the non-radiant factors are dominant, to the point where interchanged radiation becomes irrelevant as in Table 6.4.2.

Secondly, even with the maximum view factor forcing effect, the warmest surface is only 253.55 K, which is less than both the clear and hatched combo's cuboids when only one side is exposed to the Sun and even less still than if the clear and hatched combos were rotated to receive more sunlight. Also, the temperatures of the cubes after separation, experiencing full view factor radiation effect are all far less than any of the temperatures shown in any of the other examples, with the one exception of the one clear cube on its own.

Thirdly the average temperatures of the system in the full view factor radiation model are all much less than in the other models. Average temperature of the system being around 207K. This is far lower than all the others.

Finally, these calculations assuming only a 1mm spacing, if the distance is increased the effects drastically reduce further still as indicated in Table 6.5.1 and full opaque radiation forcing is applied. Gases are highly transparent, and most of the atmosphere is miles away from the surface, not 1mm away.

Not exactly a convincing and compelling argument for global warming now is it? So when Frizzlers, tell you CO_2 causes global warming because of its radiation forcing effect, you know that can't possibly be the case. An atmosphere envelope's a planet, thus increasing the surface area for emission into space and the conductive, convective and latent heat transfers will overwhelm radiation transfer processes and render it irrelevant in determining steady state temperature, except as to reduce it due increased emissions out to space.

Convection, conduction and latent heat transfers are all effective heat transfer methods and can transport far more energy from place to place than radiation can.

Take home point here? CO_2 does not cause global warming in any fashion. It can only act to reduce atmospheric temperatures.

Conclusion

Some of the maths and concepts here can be tricky to follow if this is all new to you. Remember this, CONTACT, changes the picture. When one object is connected to another by a medium, be it a solid object or a fluid one like a gas or medium, non-radiation methods of heat transfer can be dominant.

Just think of the weather on Earth. If an Arctic wind blows in from the North, you can expect cold weather, similarly if tropical weather blows in from the South, you can expect it to be warm. All our energy comes from the Sun, this is the ultimate source of our energy. Objects, including gases, with high emissivity's throw away their energy readily. This results in colder temperatures. We can see with Cubes and cuboids, that increasing emissivity decreases temperatures. So if we start to add more emissive gases into our atmosphere, it would be a complete non-sense to then say that this is something that would cause a warming to occur.

The fact that it is such a small constituent part of the atmosphere renders its effect completely negligible when compared to all the other gases and vapours present.

Touching the clear cube in space would clearly freeze our hands, and so the presence of CO_2 in the atmosphere has a cooling influence, albeit an incredibly small one due to its presence in very small quantities.

Freezing is a far cry from being Frazzled to death as the Frizzler's deadly Frazzling back rays supposedly does to us. I for one am not scared of their back ray guns, their back ray maths or their back ray science and have clearly shown it to be a complete irrelevance. The supposed "greenhouse effect" just does not exist. It is time for the adults of this planet to grow out of this fancy, fad phase Frazzle fairy tale and enter into the real world and stop wasting resources trying to present solutions to a problem that doesn't even exist.

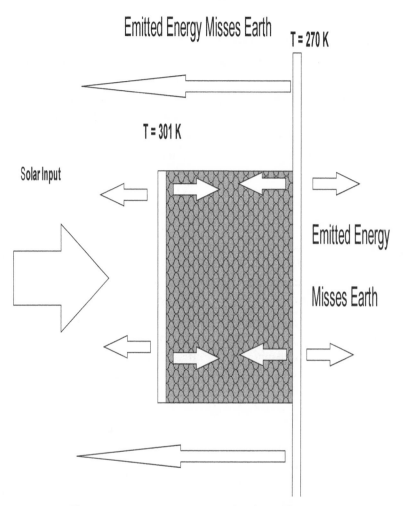

Illumination Seven—London's Calling 911

Executive Summary

United Kingdom and London authorities are always dialling 911 and telling the USA they should implement Carbon taxes and restrict CO_2 emissions, because they should "save the planet". Well, they should put that phone down and stop making "crank calls". This is a step by step guide on how to call out the 911 callers, with their "Flat Back Science."

Frizzlers, don't have a clue what they are talking about, when they tell you that back radiation from CO_2 in the air causes global warming and then tell you what you need to do is cut back, have less children, drive smaller cars, use less energy and pay more tax and just die off, evil human, you know, because your presence is just such a pain for liberal elites to deal with. It's all a con, a scam and nothing else.

Don't ever pay money for a CO_2 reduction device and don't pay any Carbon Taxes, refuse. Go on protest, send letters to your energy company, send letters to politician, tell your school to stop teaching lies to your children.

In this illumination I will use a straightforward example of calculating the view factor radiation effects of two flat plates in space. And then show how the simple addition of a transfer medium completely ruins this as the medium acts as an "equalizing agent" which nullifies all view factor effects within the system. Then I show that the atmosphere is in effect a much bigger back plate, which means much of the energy emitted from the Atmosphere, doesn't hit the Earth.

In this Illumination I will show:

1. The temperature of a flat plate in space radiating to two sides
2. What happens if I add a plate behind the first front plate.
3. I show a table of calculations which show this.
4. What happens if we add a transfer medium between the two, like water or air to create a "system"
5. You will see that "the system of two flat plates" can maintain temperatures exactly as with just the one two sided flat plate.
6. Show what happens when the back plate is larger
7. Explain how the atmosphere is three dimensional, has more surface area and is essentially a massive back plate.

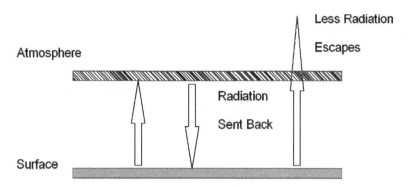

Surface warmed because of Re-emitted atmospheric radiation

Diagram 7.1

Two Line Radiation Model

Have you ever seen a diagram like this before, when people are explaining the warming effects of back radiation to you?

What do you say to someone when they show you something this, how can it possibly be wrong? Well for a start, it assumes that the Earth is flat, oh yes it does. Look at it, two flat lines with two directions of energy travel. The Frizzlers are flat Earthers.

The Earth is three dimensional and this has a massive effect on what is going on.

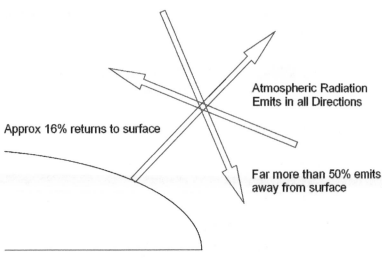

Diagram 7.2

Three Dimensional Radiation For Atmosphere

Is anyone really going to argue that 50% of the radiation is coming back? What a daft argument to make. When something is in the atmosphere there is nothing blocking the radiation going left, right, up, backwards or forwards, only some of the radiation on the way down can get re-absorbed by the surface. So let's see what potential effects this could have. We can see just by looking out the window that we should use a low back radiance 3D approach and ignore all simplistic high back radiation 2D approaches.

A Single Two Sided Flat Plate in Space

Using the steady state equation we can easily determine the temperature of a two sided flat plate in space.

$$0 = \acute{\alpha}\, E\, A(a) - \acute{\epsilon}\, A\, (e)\, \sigma\, T^4$$

If the Solar Constant of emission of energy $E = 1367\ w/m^2$ and the flat plate is a simple $1m^2$ x $1m^2$ with an A / E Ratio of 1 then the two sided plate will have a temperature of 331.35 K. This flat plate will emit energy at a rate of 683.5 w / m^2 each side.

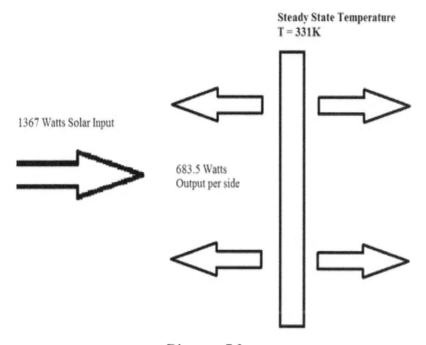

Diagram 7.3

A Two Sided Flat Plate in Space

Now if another flat plate was added behind the first plate, then we can determine what view factor effect this will have on the first plate, if any.

Two Flat Plates Back-to-Back Seperated by 1mm

So what, affect does the addition of a second plate have to the heat loss rate of the first plate?

You can see from the tables 7.2.1 & 7.2.2 & Diagram 7.2 in the following pages that the temperature of the front plate has increased from T = 331.35 K to T = 356.06 K and that the rate of heat emission from the front plate has increased from 683.5 watts to 911 watts.

The back plate, when fully shaded from the Sun by the front plate and having only the front plate as its source of heat emission maintains a temperature of 299.41 K and gives an output of 456 watts. This matches the heat input into the first plate of 1367 watts. 911 emission from the front plate and 456 emission from the front plate equals the 1367 total in.

Performing a radiation exchange check using the following formulae to ensure accuracy.

$$Q = e \, \sigma \, (T_h^{\,4} - T_c^{\,4})$$

$$Q = 1 \times 5.67E\text{-}08 \times (16{,}072{,}844{,}758 - 8{,}036{,}467{,}728)$$

$$Q = 1 \times 5.67E\text{-}08 \times (8{,}036{,}377{,}030)$$

$Q = 455.66$ Watts which rounded up is 456 Watts which exactly matches the rates on the schedules and drawings. Therefore I can know with absolute certainty that there are no errors in these calculations.

Black Dragon

Iteration No	Total Heat Input by the Sun (Watts)	Total Heat Input by the Emissions of the Second Plate (Watts)	Total of Inputs	Area of Two Sided Plate (m²) A2	Emissivity of First Plate, e	Rate of Heat Output, w/m²	Emissions out into Space, Watts	Emission out towards Second Plate, Watts	Emission Total Output	Q	e	σ	T^{-4}, $T4 = Q/e\sigma$	T
1	1,367	0	1,367	2	1	684	684	684	1,367	683.5	1	5.67E-08	12,054,673,721	331.35
2	1,367	342	1,709	2	1	854	854	854	1,709	854.38	1	5.67E-08	15,068,342,152	350.36
3	1,367	427	1,794	2	1	897	897	897	1,794	897.09	1	5.67E-08	15,821,759,259	354.66
4	1,367	449	1,816	2	1	908	908	908	1,816	907.77	1	5.67E-08	16,010,113,535	355.71
5	1,367	454	1,821	2	1	910	910	910	1,821	910.44	1	5.67E-08	16,057,202,105	355.97
6	1,367	455	1,822	2	1	911	911	911	1,822	911.11	1	5.67E-08	16,068,974,248	356.04
7	1,367	455	1,823	2	1	911	911	911	1,823	911.28	1	5.67E-08	16,071,917,283	356.05
8	1,367	455	1,823	2	1	911	911	911	1,823	911.32	1	5.67E-08	16,072,653,042	356.06
9	1,367	455	1,823	2	1	911	911	911	1,823	911.33	1	5.67E-08	16,072,836,982	356.06
10	1,367	455	1,823	2	1	911	911	911	1,823	911.33	1	5.67E-08	16,072,882,967	356.06
11	1,367	455	1,823	2	1	911	911	911	1,823	911.33	1	5.67E-08	16,072,894,463	356.06
12	1,367	455	1,823	2	1	911	911	911	1,823	911.33	1	5.67E-08	16,072,897,337	356.06
13	1,367	455	1,823	2	1	911	911	911	1,823	911.33	1	5.67E-08	16,072,898,055	356.06
14	1,367	455	1,823	2	1	911	911	911	1,823	911.33	1	5.67E-08	16,072,898,235	356.06
15	1,367	455	1,823	2	1	911	911	911	1,823	911.33	1	5.67E-08	16,072,898,280	356.06

Table 7.4.1

Absorption and Emission From Front Plate One

101

	Absorption Side						Emission Side					
Iteration No	Total Heat Input by the Emissions of the First Plate (Watts)	Area of Second Plate (m²)	Emissivity of Second Plate	Rate of Heat Output	Emissions out into Space	Emissions back towards First Plate	Total Output	$Q = e\,\sigma\,T^4$	e	s	$T^4 = Q/e\,\sigma$	T
		A2	a	w/m²	Watts	Watts		Q	e	s	T^4	T
1	684	2	1	342	342	342	684	341.75	1	5.7E-08	6,027,336,361	278.63
2	854	2	1	427	427	427	854	427.19	1	5.7E-08	7,534,171,706	294.52
3	897	2	1	449	449	449	897	448.55	1	5.7E-08	7,910,879,600	296.23
4	908	2	1	454	454	454	908	453.89	1	5.7E-08	8,006,636,768	296.12
5	910	2	1	455	455	455	910	455.22	1	5.7E-08	8,020,601,253	296.54
6	911	2	1	456	456	456	911	455.55	1	5.7E-08	8,034,487,124	296.59
7	911	2	1	456	456	456	911	455.64	1	5.7E-08	8,036,958,642	296.4
8	911	2	1	456	456	456	911	455.66	1	5.7E-08	8,038,326,521	296.4
9	911	2	1	456	456	456	911	455.66	1	5.7E-08	8,038,418,491	296.4
10	911	2	1	456	456	456	911	455.67	1	5.7E-08	8,038,441,483	296.4
11	911	2	1	456	456	456	911	455.67	1	5.7E-08	8,038,447,222	296.4
12	911	2	1	456	456	456	911	455.67	1	5.7E-08	8,038,448,669	296.4
13	911	2	1	456	456	456	911	455.67	1	5.7E-08	8,038,449,228	296.4
14	911	2	1	456	456	456	911	455.67	1	5.7E-08	8,038,449,118	296.4
15	911	2	1	456	456	456	911	455.67	1	5.7E-08	8,038,449,140	296.4

Table 7.4.2

Absorption and Emission From Back Plate Two

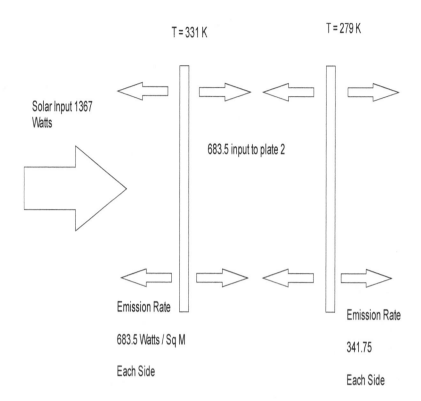

T = 331 K   T = 279 K

Solar Input 1367 Watts

683.5 input to plate 2

Emission Rate

683.5 Watts / Sq M

Each Side

Emission Rate

341.75

Each Side

Total Out = 683.5 + 683.5 = 1367 Watts from Object 1

Diagram 7.4.3

Front and Back Plates Seperated by 1mm in Space Radiating to Each Other

It can be seen from the above calculations, table and drawings that the figure of output from the front plate has increased from 683.5 to 911 w / m^2 hence the title of this illumination.

This is the sort of proof which "Frizzlers" worldwide would like to believe is what is causing global warming. How wrong they are. This is a 2 dimensional model, and we already know that the two-dimensional model can't be right.

The authorities of the UK and London are always calling USA and telling them that they should restrict their carbon emissions, in order to prevent climate change telling them they are so irresponsible if they do not. The flat Earth 911'ers and Frizzlers should all get a clue and learn what is really going on instead of spouting off nonsense to get a good applause, to look magnanimous and to push forth their progressive high taxation socialist and communist one world order government agenda.

However, as I showed in Illumination Five, these types of calculations only work in space, in an environment where there is a pure Vacuum. Also don't forget the plates are fully opaque to IR and are extremely close together to maximise forcing.

If there is a liquid or gaseous medium, surrounding, joining or binding the objects then the heat transfer processes of conduction, convection and latent heat transfers nullify all view factor effects as they act as bridges to transfer heat energy. Just as with the filament of a light bulb, the gas acts to cool the filament and warm the glass bulb.

Now in section 7.5 onwards I will go onto explain how other process can effect these temperature profiles. Before I do so, you should be aware that there exists the possibility that this 911 maths is factually inaccurate, because it assumes that back-radiance as a process is real and can occur. What if it was not real and did not occur?

Then instead of an increase in temperature of the first plate rising from 331K to 356K we instead find that the temperature of the first plate remains at 331 K and that the 2nd plate has a temperature of 279 K this would be because it is absorbing 684 Watts of energy and can therefore only emit 684 Watts of energy and it would do this in two directions. What this method would assume is that the energy going from the 2nd plate towards the 1st plate, is simply just not absorbed by the firs plate as it is already at its blackbody maxima at that temperature.

I have no way of determining if this is correct or not without appropriate laboratory facilities. One simple experiment could be constructed with plates within a vacuum to see if this true or not, but unfortunately I find no records of such an experiment ever been performed nor have I witnessed one. If an experiment was performed with plates in a vacuum that back radiance had no effect at all on the temperature on the first plate then the following explanations in 7.5 onwards, are unnecessary as this one item on its own would show all back radiation maths to be a complete sham. As it has never been done before, I am therefore going to move forward assuming that such back radiant maths is correct and move forward on that basis and show how radiation can be rendered irrelevant by other forces even when back-radiance maths is applied to its fullest.

If Back Radiance was found to be false, a two plate model might look like this in 7.4a where there is no input into the first plate from the 2nd plate at all.

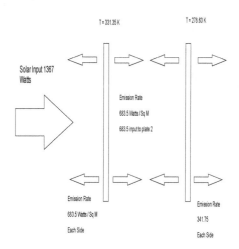

Total Out = 683.5 + 683.5 = 1367 Watts from Object 1

Diagram 7.4a
No Back Radiance Absorption from the 2nd Plate At All

If it was found in a lab that this was how it was, then clearly it would be "game over" for all back radiation greenhouse effect frazzlers. In the absence of proof of either answer, we can move forward on the basis of "as if it was correct" so let's do that.

Adding a Heat Transfer Medium to the Mix—Conductance

If I added a heat transfer medium such as air or water between the two plates, then what happens is the "system" starts to equalize temperatures between the two plates. This is because as heat transfers from the front plate, it reduces the temperature of the front plate and as the heat transfer medium warms, it becomes possible for the medium to transfer its heat to the second plate. The medium, is adding an additional path of heat transfer from the first plate to the second plate, increasing the rate of heat loss from the first and acting as a heat source to the second. This has the effect of warming the second plate.

Given enough time, and enough medium this will cause equilibrium to occur.

The rate of heat transfer because of conductance is calculated as follows.

$Q / t = (k A (T^1 - T^2)) / d$ whereby :-

k = Thermal Conductivity.

D = Distance of material.

A = Area of material.

T^1 = Temperature of First Barrier.

T^2 = Temperature of Second Barrier.

So in two plates example, if I was to add air, then the rate of heat transfer would be as follows.

K air = 0.024 w / m k d = 0.001m A = 1 T^1 = 356.06 & T^2 = 299.41.

Q = (0.024 * 1 * (56.65)) / 0.001 = 1359.6 Watts of energy transfer from the plate to the air. This very high rate of heat transfer is far in excess of even the radiation emission from the surface given at these temperatures.

Therefore, the heat conduction process becomes the dominant factor in heat transfer.

This is the reason that thermos flasks work better when they utilise a vacuum instead of air. The air acts as a bridge between the two barriers and increases the rate of heat transfer between the two surfaces. Having a vacuum reduces the rate of heat transfer and enables the contents of the flask to stay warmer for longer.

The presence of air enables heat to transfer from the first plate to the second plate with ease. Because molecules of air bounce into the plate and then move across to other side and heat the second plate, transferring energy as they do so.

So if we introduced thermal heat capacity into the mix, to try to figure out how long before the temperatures roughly equalised we would find the following happens.

In Table 7.5.1 we can see that the temperature of the first plate drops, which reduces the amount of radiation being emitted to the back plate and the amount of radiation being emitted out into space increases from the back plate. The amount of energy transferred reduces eventually (after just 1490 seconds in this example assuming plate heat capacities of 1000 KJ kg^{-1} k^{-1}.) until a temperature differential between the plates of only 19 K exists. With eventual temperatures of 340.36 K for the front plate and 321.55 K for the back plate with overall energy in matching energy out at 1367 watts.

If we had a transfer medium with a higher conductance this equalisation process would improve. For example if the conductivity was increased to 0.10 W m-^{1}K^{-1} then the temperature differential drops to 6 degrees K and with a thermal conductivity of 0.99 W m-1 K-1 it drops to zero. The temperature of both plates at that level of conductivity exactly matches the one plate on its own, T = 331.35K.

This equalisation happens because the rate of heat out of the Front Plate 1from non-radiation means increases. Once it matches the rate of heat in as a result of solar radiation the level at which heat enters the system, then the rate of heat transference from the front plate can match the rate at which heat enters in from the front plate, it is effectively transferred straight to the back plate through the transfer medium. This heats up the back plate until it matches the same temperature of the front plate, which reduces in temperature so that the rate of heat out of the system matches the rate of heat into the system.

The above, of course, assumes no heat losses from the air between the plates would occur, which of course it would, reducing overall system temperatures.

We do not need to be reliant on just air conductance. Convection is an effective means of heat transfer also.

Insulation No	Specific Heat Capacity of Plate	Solar In (Space)	Q Plate out (Space)	Temp Front Plate	Q Plate Out (Internal)	Energy Stored in Front Plate (Internal)	Q Plate Out (Internal)	Temp Back Plate	Q Plate Out (Space)	Energy Stored in Back Plate	Temp Diff	Q Rate of Thermal Space	Total Space
1	1,000.00	1,357.00	911.33	356.06	911.33	356,060	455.67	299.41	455.67	299,410	57	1,559.60	1,567.30
2	1,000.00	1,357.00	897.49	354.7	897.49	354,700	463.92	300.76	463.92	300,756	54	1,284.67	1,561.41
3	1,000.00	1,357.00	884.82	353.44	884.82	353,442	471.69	302.01	471.69	302,007	51	1,234.42	1,558.50
4	1,000.00	1,357.00	873.21	352.38	873.21	352,276	479	303.17	479	303,172	49	1,178.51	1,552.21
5	1,000.00	1,357.00	862.56	351.2	862.56	351,197	485.88	304.25	485.88	304,255	47	1,126.62	1,548.44
6	1,000.00	1,357.00	852.79	350.2	852.79	350,199	492.35	305.36	492.35	305,252	45	1,078.47	1,545.14
7	1,000.00	1,357.00	843.82	349.27	843.82	349,274	498.43	306.2	498.43	306,200	43	1,033.77	1,542.74
8	1,000.00	1,357.00	835.57	348.42	835.57	348,418	504.13	307.07	504.13	307,072	41	990.29	1,539.71
9	1,000.00	1,357.00	828	347.53	828	347,626	509.49	307.88	509.49	307,884	40	953.79	1,537.49
10	1,000.00	1,357.00	821.03	346.69	821.03	346,892	514.51	308.54	514.51	308,640	38	918.05	1,535.54
11	1,000.00	1,357.00	814.63	346.21	814.63	346,214	519.22	309.34	519.22	309,344	37	884.87	1,533.85
12	1,000.00	1,357.00	808.74	345.59	808.74	345,586	523.63	310	523.63	309,999	36	854.08	1,532.37
13	1,000.00	1,357.00	803.31	345	803.31	345,005	527.77	310.61	527.77	310,609	34	825.49	1,531.08
14	1,000.00	1,357.00	798.32	344.47	798.32	344,457	531.64	311.18	531.64	311,173	33	798.96	1,529.96
15	1,000.00	1,357.00	793.72	343.97	793.72	343,971	535.27	311.71	535.27	311,707	32	774.33	1,328.39
1489	1,000.00	1,357.00	760.87	340.36	760.87	340,355	606.13	321.55	606.13	321,548	19	451.38	1,567.00
1490	1,000.00	1,357.00	760.87	340.36	760.87	340,355	606.13	321.55	606.13	321,548	19	451.38	1,567.30

Table 7.5.1 How Conductance of air Modifies Temperatures of the Plates Whilst Still Showing View Factor Radiance

109

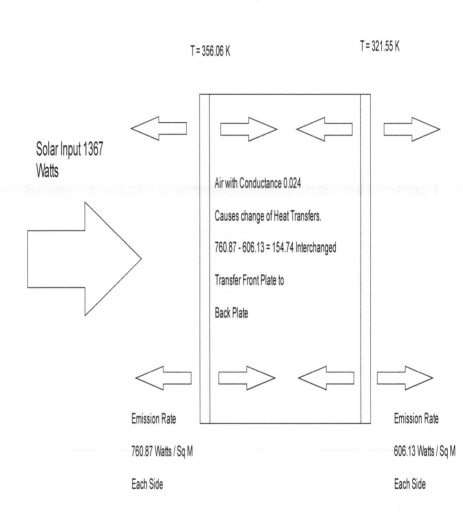

T = 356.06 K   T = 321.55 K

Solar Input 1367 Watts

Air with Conductance 0.024

Causes change of Heat Transfers.

760.87 - 606.13 = 154.74 Interchanged

Transfer Front Plate to

Back Plate

Emission Rate

760.87 Watts / Sq M

Each Side

Emission Rate

606.13 Watts / Sq M

Each Side

Total Out = 760.87 + 606.13 = 1367 Watts

Diagram 7.5 Two Plates with Air Allowing for Heat Conductance Transfer

Convection in the Heat Transference Medium

Convection is a highly effective mechanism for heat transfer, especially for liquids and gases. The movement of the molecules in the air allows for them to bash into each other and thus cause a transfer of energy when the molecules have temperature differences. The convection heat transfer equation is as follows:-

$$Q = h_c A (T^1 - T^2) \text{ whereby:-}$$

h_c = convective heat transfer co-efficient.

A = Area.
T^1 and T^2 are the temperatures of the two barriers.

The heat transfer co-efficient varies not only with the type of gas but also on the velocity of the gas. The higher the velocity the higher this heat transfer co-efficient will be.

For air this is. $Hc = 10.45 - v + 10\ v^{1/2}$

So in our example, if I assumed 1mm was enough space to start up a convection current, (which it would not be, the frictional forces would prevent it in such a small space) and then also assumed that the plates were still and air had no velocity to speak off, we would have a convective co-efficient of air of just 10.45.

This would result in a convective heat transfer into the air of = 10.45 * 56.65 = 591.993 Watts. Now a point to note is, on the Earth the rate of heat transfer as a result of convection is much higher than this, because air speeds are often not still but can be 20, 30, 50 mph or even more. This would have a massive multiplier effect on the amount of energy transferred because of convection into the atmosphere.

So let's see what happens if we add Convection to the table.

We see that after 1500 seconds the difference levels at 15k between the two plates is 15 K, which for the first plate at 338.41K is a mere 7.06 K higher than the plate on its own. This still assumes that there are no losses at all from the gas, which let's not forget is itself at an equivalent temperature. In space, this would be emitting out energy in all directions, up, down, to the sides and these energy losses, which are acting to increase surface emission area of the system would cause temperatures to drop.

On Earth a great deal of the heat which enters the atmosphere is as a result of convective heat transfer from the surface, upwards. It so powerful it creates a bulge in the atmosphere every day known as the "Di-urnal atmospheric bulge" This bulge is pulled in two ways, one towards the moon due to its gravity and another towards the sunlight each day. The one towards the Sunlight occurs because the atmosphere is absorbing energy directly, but also because as the air heats up it starts up very powerful convection currents which push the atmosphere upwards several miles higher than it the stratified level in the night. Frizzlers like to pretend this effect is minimal. It is not.

Iteration No	Front Plate						Back Plate							
	Specific Heat Capacity of Plate	Solar In	Plate out (Space)	Temp Front Plate	Plate Out (internal) Plate	Energy Stored in Front (internal)	Plate Out (internal)	Temp Back Plate	Plate Out (Space)	Energy Stored in Back Plate	Temp Diff	Q Rate of Thermal Conductance	Total Space	Q Convection energy transfer
1	1,000.00	1,367.00	911.33	356.06	911.33	356.050	455.67	299.41	455.67	299.410	57	1,359.60	1,367.00	591.99
2	1,000.00	1,367.00	891.51	354.11	891.51	354.108	467.54	301.34	467.54	301.342	53	1,266.40	1,359.06	551.41
3	1,000.00	1,367.00	873.86	352.34	873.86	352.342	478.54	303.1	478.54	303.098	49	1,181.65	1,352.40	514.6
4	1,000.00	1,367.00	858.11	350.74	858.11	350.745	488.71	304.7	488.71	304.696	46	1,105.14	1,346.81	481.2
5	1,000.00	1,367.00	844.04	349.3	844.04	349.297	498.1	306.15	498.1	306.149	43	1,035.53	1,342.13	450.9
6	1,000.00	1,367.00	831.45	347.99	831.45	347.987	506.75	307.47	506.75	307.470	41	972.4	1,338.20	423.4
7	1,000.00	1,367.00	820.18	346.8	820.18	346.802	514.73	308.67	514.73	308.673	38	915.1	1,334.91	398.45
8	1,000.00	1,367.00	810.09	345.73	810.09	345.730	522.06	309.77	522.06	309.767	36	863.11	1,332.15	375.91
9	1,000.00	1,367.00	801.03	344.76	801.03	344.760	528.81	310.76	528.81	310.763	34	815.93	1,339.84	355.27
10	1,000.00	1,367.00	792.91	343.88	792.91	343.885	535.01	311.67	535.01	311.669	32	773.11	1,327.92	336.65
11	1,000.00	1,367.00	785.62	343.09	785.62	343.089	540.7	312.49	540.7	312.495	31	734.25	1,326.31	319.71
12	1,000.00	1,367.00	779.06	342.37	779.06	342.372	545.92	313.23	545.92	313.246	29	699	1,324.98	304.36
13	1,000.00	1,367.00	773.16	341.72	773.18	341.725	550.71	313.92	550.71	313.921	28	667	1,323.89	293.42
14	1,000.00	1,367.00	767.89	341.14	767.89	341.137	555.1	314.56	555.1	314.535	27	637.96	1,322.99	277.78
15	1,000.00	1,367.00	763.13	340.61	763.13	340.607	559.12	315.12	559.12	315.124	25	611.61	1,322.26	266.3
1489	1,000.00	1,367.00	743.64	338.41	743.64	338.412	623.36	325.81	623.36	325.809	15	350.47	1,367.00	152.6
1490	1,000.00	1,367.00	743.64	338.41	743.64	338.412	623.36	325.81	623.36	325.809	15	350.47	1,367.00	152.6

Table 7.6.1 Convection Effects Added

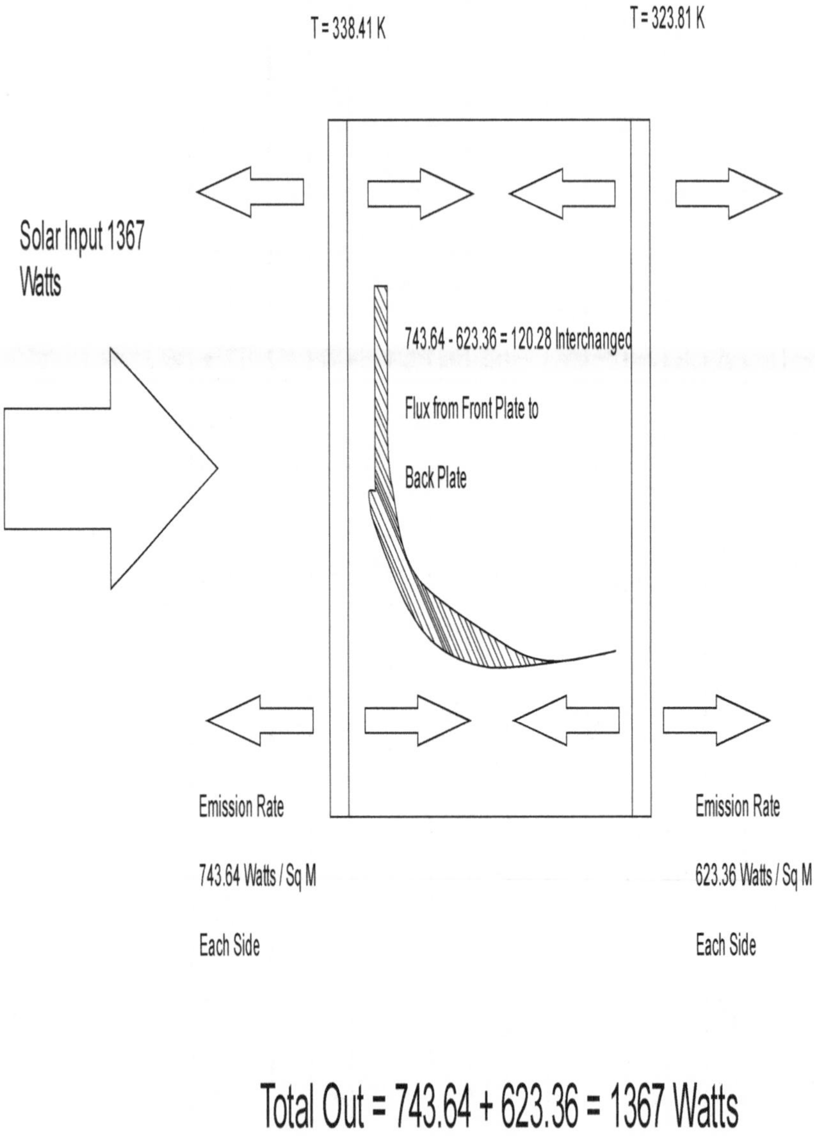

T = 338.41 K T = 323.81 K

Solar Input 1367
Watts

743.64 - 623.36 = 120.28 Interchanged

Flux from Front Plate to

Back Plate

Emission Rate

743.64 Watts / Sq M

Each Side

Emission Rate

623.36 Watts / Sq M

Each Side

Total Out = 743.64 + 623.36 = 1367 Watts

Diagram 7.6.2 Convection Effects Added to Heat Transfer Process

Convection is an extremely potent force on Earth.

"In each hemisphere there are three cells (Hadley cell, Ferrel cell and Polar cell) in which air circulates through the entire depth of the troposphere. The troposphere is the name given to the vertical extent of the atmosphere from the surface, right up to between 10 and 15 km high. It is the part of the atmosphere where most of the weather takes place." [35]

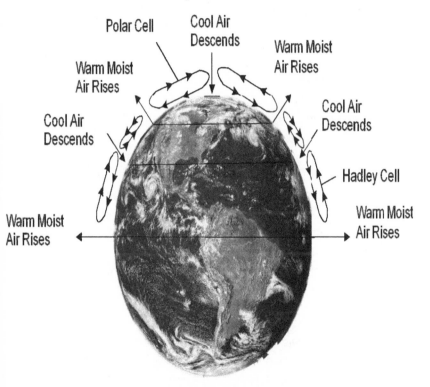

Diagram 7.7 Hadley Cells Transfer Heat Around the Earth Via Convection

When water evaporates, sunlight energy is converted into latent energy in the water. This causes the water to vaporise into the air and remains inside the water vapour.

---

[35] Metoffice, 2017

When the water vapour converts back into water droplets, this latent heat energy is converted back into heat which provides a warming effect to the air around it as it does so.

This has the effect of transferring heat from the surface and to the troposphere, which is another heat transfer process which could be used in the plate to plate example to reduce the temperature difference further still.

I could add another column to the table and another diagram, however, I believe you will understand what is occurring now.

The more methods there are for transferring heat from one side to the other, the more and more any temperature difference which exists is offset.

Atmosphere Increases the Surface Area for Emission

The atmosphere sits on top of the planet. This in effect means that the planetary atmosphere has a larger emission area than the surface of the Earth.

This has a different radiant forcing effect, because instead of being akin to two plates of equal size, they are now instead akin to two plates of different size, with the back plate, being larger than the front plate. This means that the back plate is going to be a much more effective radiator and because not all of its radiation comes back towards the front plate, some is lost out to space.

So instead of being something which causes the front plate to rise in temperature, when the heat transfer medium is added, it becomes something which reduces the temperature of the front plate.

Now if we go back to the plate to plate example, what happens if we increase the size of the back plate to reflect this increase in emission area, which the atmosphere represents?

Well, what we find is that the amount of energy coming back from the back plate, is less, this is because more of the back plate is exposed to space.

See Diagram 7.6.

In this example I have of course assumed that the back plate won't be in receipt of any energy from the sun. This is because I am attempting to simulate the effects of Carbon Dioxide and other emissive gases, which do not absorb sunlight, and so such energy is ignored.

In the real world, of course the rate at which the atmosphere absorbs sunlight will act to increase the absorption rate and therefore the rate at which energy is absorbed. And this is direct absorption of sunlight, in the much larger atmosphere is a great contributor to heating of the Earth, especially as the atmosphere is mostly composed of low emissive gases. Just like the low emissive Cuboids, they experience the highest temperatures when exposed to sunlight.

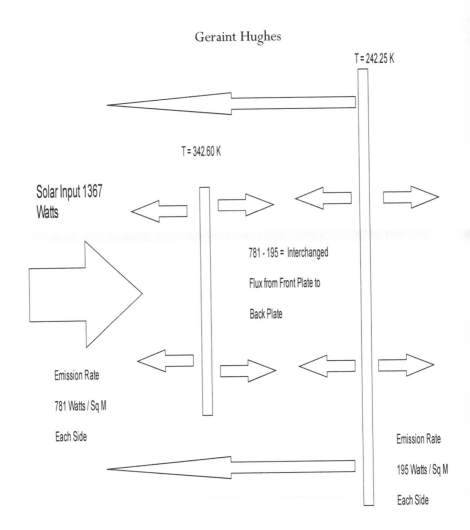

Total Out = 781 + 586 = 1367 Watts

7.9.1 Atmosphere Acting as a Bigger Back Plate Because it is a Larger Sphere than the Planet

And what if I add air for conductivity effect?

Iteration No	Specific Heat Capacity of Plate	Front Plate					Back Plate						
		Solar in (Space)	Temp Plate out (Space)	Temp Front Plate	Plate Out (Internal) Front Plate	Energy Stored in Front Plate (Internal)	Plate Out (Internal)	Temp Back Plate	Plate Out (Space)	Energy Stored in Back Plate	Temp Diff	Q Rate of Thermal Conductance	Total Space
1	1,000.00	1,367.00	781.15	342.6	781.15	342,600	195.27	242.25	585.81	242,250	100	2,468.40	1,366.95
2	1,000.00	1,367.00	759.41	340.19	759.41	340,192	203.08	244.64	609.24	244,637	96	2,293.32	1,368.66
3	1,000.00	1,367.00	739.59	337.95	739.59	337,950	210.56	246.85	631.67	246,857	91	2,186.21	1,371.25
4	1,000.00	1,367.00	721.48	335.86	721.48	335,862	217.69	248.92	653.07	248,923	87	2,085.53	1,374.55
5	1,000.00	1,367.00	704.91	333.92	704.91	333,917	224.49	250.84	673.47	250,845	83	1,993.76	1,378.38
6	1,000.00	1,367.00	689.74	332.1	689.74	332,105	230.95	252.63	692.85	252,629	79	1,907.42	1,382.58
7	1,000.00	1,367.00	675.81	330.42	675.81	330,416	237.08	254.29	711.23	254,288	75	1,827.06	1,387.04
8	1,000.00	1,367.00	663.02	328.84	663.02	328,841	242.88	255.83	728.64	255,830	73	1,753.26	1,391.66
9	1,000.00	1,367.00	651.26	327.57	651.26	327,373	248.36	257.25	745.09	257,262	70	1,682.65	1,396.35
10	1,000.00	1,367.00	640.43	325	640.43	326,003	253.54	258.59	760.61	258,592	67	1,617.87	1,401.04
11	1,000.00	1,367.00	630.44	324.72	630.44	324,725	258.41	259.83	775.24	259,826	65	1,557.57	1,405.68
12	1,000.00	1,367.00	621.23	323.53	621.23	323,532	263	260.97	789	260,971	63	1,501.45	1,410.22
13	1,000.00	1,367.00	612.72	322.42	612.72	322,418	267.31	262.03	801.92	262,032	60	1,449.25	1,414.64
14	1,000.00	1,367.00	604.85	321.38	604.85	321,377	271.35	263.02	814.04	263,018	58	1,400.62	1,418.89
15	1,000.00	1,367.00	597.56	320.41	597.56	320,406	275.13	263.93	825.4	263,931	55	1,355.39	1,422.95
1,489	1,000.00	1,367.00	463.97	300.76	463.97	300,765	301.01	266.93	903.03	269,929	31	740.07	1,367.00
1,490	1,000.00	1,367.00	463.97	300.76	463.97	300,765	301.01	266.93	903.03	269,929	31	740.07	1,367.00

Table 7.9.1 Table with Larger Plate Including Air
Conductance Transfer

119

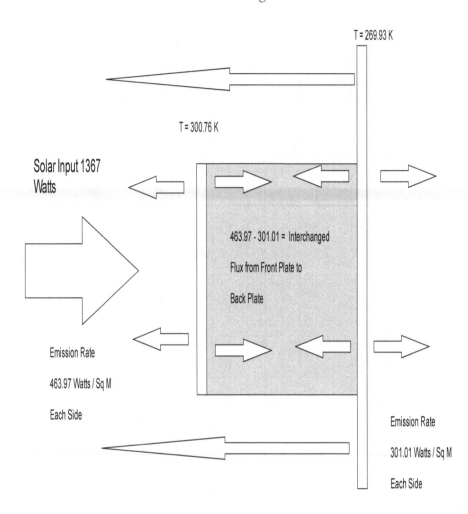

Total Out = 463.97 + 903.03 = 1367 Watts

Diagram 7.9.2 Larger Back Plate with Air Conductance

Here you can see that the back plate, suddenly becomes responsible for a greater proportion of the emissions out into space, despite holding lower temperatures.

There is still a difference in temperatures, with the front plate being warmer than the back plate, but the front plate is now much cooler than the first 2 sided example which the 911 callers love to use when explaining "so called greenhouse effect".

You can see that in this example, the temperature difference is 31deg K.

What happens to the temperatures if I include a small amount of convection?

The first surface cools further still as more of its energy is transferred into the back plate. The front plate maintains a temperature of 295.71 K with the back plate being 272.17 K. And it would be cooler further still if I added in large amount of convection and large amounts of latent heat transfer.

We now are arriving at an explanation which shows how things on Earth are. With a warm surface and a cooler troposphere and Stratosphere, emitting a greater proportion of the energy into space, with large proportions of that energy being inputted as a result of convection and latent heat transfers.

The big reason for this change in relationship between the back-plate being responsible for more than 50% of the system emissions, is what swings it to being a big radiator. Now, the surface of the Earth at any point can only emit one way, upwards and out into space. Items in the atmosphere however can emit in all directions. And that includes to the sides. A cube shaped object in the troposphere, will be emitting not 50% of its radiation down to the surface but something more along the lines mere 16% with the remaining 83% going out to space. All objects in the atmosphere, including gases and clouds are all doing the same thing. This means all of the atmosphere is acting like a much bigger back plate.

Atmosphere: More Volume and Surface Area than the Earth

Now, is the back plate of the atmosphere really bigger than the front plate? Well yes indeed it is.

If I assumed that the Earths diameter was 12,756 km then the surface area of the Earth would be 511,185,933 km^2. If I assumed the troposphere was just an additional 17km, then the surface area is 513,914,606.80 km^2. This represents an increase in surface area of almost 3 million square kilometres. At 2,728,674.28 km^2. That's in terms of equivalence that's like missing out of all of India. If I go out to the stratosphere with 50km height, this difference increases to over 10 million square kilometres equivalent to Brazil and if I go right out to the Earths Karman line, which represents the boundary of the atmosphere of 100km the difference increase further still to 27 million square kilometres which is more than Russia.

The Earth and its atmosphere is "3 DIMENSIONAL." What this means is that it will not only be increasing its losses to atmosphere by missing part of the Earth by reflecting backwards, but it is also missing the Earth by reflecting sideways and up and down also as mentioned early on in this Illumination. This prevents the two dimensional flat two plate 911 model from being correct.

What this means is the back plate, is less of a plate and more like a cube. When we add in the heat transferring effects of conductance, convection and latent heat you can see how increasing the emissivity of the atmosphere can cause big drops in temperature. And this is without factoring any reductions for Infrared reflectivity of the surface. On top of all that my back plate diagrams and tables, assumes full black body surfaces. The view factor radiation effects are at their maximum.

If you find this hard to comprehend, look at these plan and sectional drawings in Diagram 7.10.1.

SECTIONAL VIEW

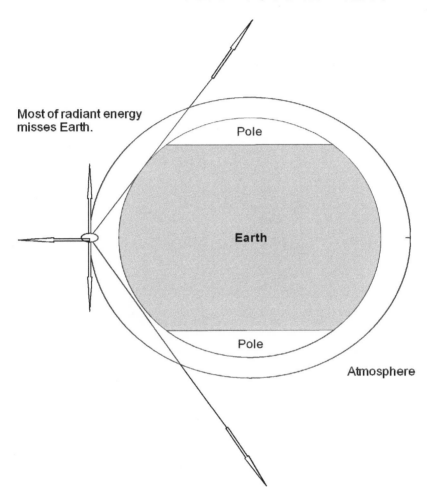

Diagram 7.10.1

**View of Earth Sectional to Illustrate Three-Dimensional
Nature of Atmospheric Emissions**

PLAN VIEW

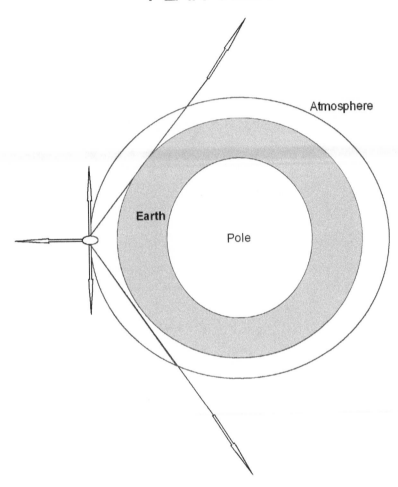

Diagram 7.10.2

View of Earth Plan to Illustrate Three-Dimensional Nature of Atmospheric Emissions

Conclusion

The instant something is in the atmosphere, when looked at from a 3 dimensional point of view more than 50% of an object's emissions go to space. If a cube was in the atmosphere, one side would absorb and exchange with the surface, but the other 5 sides would be emitting to space. That would be a return rate of 16 % not 50% and as the cube got higher up the atmosphere then its return rate would keep reducing. If the object was an emissive object, i.e. it had an emissivity higher than the air which surrounds it, it would exert a big cooling effect on the atmosphere and the conduction, convection and latent heat transfers into it would increase that effect.

What this means, is that all of the atmosphere is acting as a bigger back plate even at the lower portions of the Troposphere because much more than 50% of energy exchanged into the atmosphere is being emitted out of the atmosphere..

Yet all the Frizzler back radiation drawings, show one or sometimes two lines, with some 50% reflection back. Just completely and utterly ridiculous.

This three dimensional view of irradiance, explains why there is a temperature difference between the atmosphere and the surface, with the surface being warmer and the atmosphere being cooler, whilst simultaneously the atmosphere is responsible for a greater proportion of emission out to space than the surface meaning that increasing the emissivity of the atmosphere decreases surface temperatures.

Frizzler maths, is clearly some sort of bizarre "Flat Earth" model and is completely idiotic to try to use it to resolve 3 dimensional maths let alone base public taxation policy upon it. Yet the "Frizzlers" go around calling everyone who disagree with them "Deniers and Flat Earthers." It is they which are the Flat Earthers and they whom are the deniers.

The Frizzlers, should be ignored and policy based upon sound facts and science.

"Radiation Greenhouse Effect" is a complete nonsense as I have demonstrated and all of climate science which basis itself upon this flawed 2 dimensional Flat Earth 911 concept should be scrapped. All Carbon taxes should be scrapped and all CO_2 control measures repealed and ignored.

It's time to call out, the 911 callers and their silly "Frizzle Frazzle".

Oil is not a Fossil Fuel

Executive Summary

OIL IS NOT a fossil fuel, it does not come from decomposed remains of long dead dinosaurs, it is a fully renewable resource, which is created by diatomic algae and is therefore not something that will run out but is something that can be sustained for all time as long as the sun shines.

This Chapter will briefly cover an area of research which could actually encompass several books within itself, but I will touch on the basics of where oil actually comes and shed light on some of the most interesting research I believe which currently exists.

Fossil Fuels and the theory that they come from old rotted fossils is another example of a great mistake which the frizzle frazzler's of the left, always go on about.

They believe we are going to reach "peak oil" and that oil is a fossil fuel, which comes from the remains of long dead animals and at some point in time we will therefore use up all the oil and then there will be none left. Which of course will cause a world-wide calamity and so therefore to avoid this we all need to stop using oil and start paying carbon and other green taxes as well has having CO_2 legislation imposed upon us to restrict our use of oil

They use this argument even if it is shown that CO_2 has no effects on atmospheric temperatures and conditions at all. They can just say, "one day there will be no oil, it will be all gone, so we best get used to that situation now, rather than later. That is what a wise person would do."

Unfortunately this is just another one of those damn twaddle talking made up fantasy rubbish fairy tales, best left in the past and not a high tech forward looking high energy future.

Oil, is not a fossil fuel, it does not come from the decomposed remains of long dead mammoths, dinosaurs and fishes at the bottom of the ocean. No, not at all.

So where does it come? Well the answer is simple. It comes from Diatom algae, little molecules which absorb sunlight and then convert this sunlight into oil and these miniature critters tend to live in the seas, oceans and other watery environments and can make oil all day long throughout their entire lives. How the oil gets from these diatomic algae, to under the ground to be drilled by an oil rig is as a result of a combination of effects from the natural algae life cycle as it moves from the sea surface to the sea bed, sea and ocean circulation cycles around the world, sub surface sea environment, underground oil migration patterns as well as localised sea bed biological habitats and seabed subduction.

Figure 8.1 shows an example of what a diatomic molecule looks like.

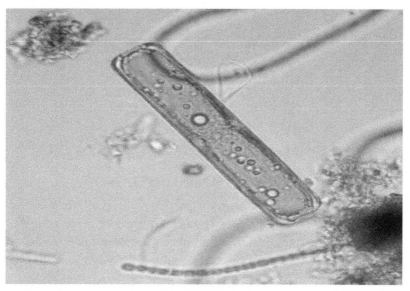

Diagram 8.1

Picture of a Diatom[36]

Diatoms have been known about since the 1700's as microscopes were first invented and people could see them and describe for the first time. In the early part of the 20th Century, the petroleum industry became interested in them as their oil producing capacity was recognised.

The National Renewable Energy Laboratory (NREL) facility, of the US Department of Energy (DOE) instigated research into which type of algae / plant life could produce the most oil as an alternative source of fuel from oil from the ground. This research is known as the Aquatic Species Program.

Here is a table which shows the difference in yields per acre for various plant type.

---

[36] Britannica 2018

Corn	18
Soybeans	48
Safflower	83
Sunflower	102
Rapeseed	127
Oil Palm	635
Micro Algae	5000-15000

Table 8.1 Comparison of Yields for Plant Types

Yields (Gallons of oil per acre per year)[37]

Here it can be seen that micro algae has the highest yield of the various plant types and as this plant life is present in most of the oceans of the planet, it is not at all surprising that we find ourselves living on a planet awash with sources of oil. It's simply a matter of research to discover how exactly the oil gets from the algae to sources under the ground, to the oil fields that we use today.

Without the presence of these oil organisms on Earth, there would be no oil, regardless of whether or not Dinosaurs and mammoths existed, millions of years ago.

All we need for oil production, is sunlight, water and these algae and as we have all three present in abundant amounts, we will have oil in abundant amounts.

Conducting research into the lifecycles of these organisms and how the oil gets from their bodies to the oil fields can be a tricky business and requires lots of complicated research, however people have studied Diatoms for many different purposes.

This particular study "Regulation of Intertidal Microphytobenthos Photosynthesis over a Diel Emersion Period Is Strongly Affected by Diatom Migration Patterns"

An extract from the study says "The MPB biomass in the photic zone (0–0.5 mm) of the studied intertidal sediment showed strong diel variability with sediment Chl concentrations showing a

---

[37] Oilgae.com, 2018

more than 2-fold increase and reaching maximal values of ~0.2 mg Chl a cm^{-3}. These changes can be attributed to the migration of diatom cells from deeper layers to the uppermost 0.5 mm of the sediment, allowing cells to absorb light to drive photosynthesis. A very strong attenuation of photosynthetic available light with depth was observed, and irradiance levels below 0.5 mm were <5% of incident downwelling irradiance. Migratory rhythms of the MPB populations of intertidal mudflats synchronized with diurnal and tidal cycles, commonly identified by the appearance of a golden brown color at the sediment surface, have been frequently described in the literature" and "For example, Saburova and Polikarpov (2003) observed that epipelic diatom cells in different phases of mitosis were found almost exclusively in the aphotic zone of the sediment and argued that deeper sediment layers provide more favorable nutrient conditions for cell growth and division."

What this study shows us is that Diatomic organisms which occur around coastal areas and estuaries, is that the organism rises closer up to the surface during sunlight hours to maximize photo synthetic production and then goes back down to the sea bed after sunlight hours to multiply.

Now that we know that these organism deliberately move themselves up and down with the tides and currents in order to maximize their receipt of sunlight and then go back down to the seabed to multiply we can see that it isn't such a great stretch of imagination to say that these organisms when they are on the sea bed can get drawn down by sea-bed subduction into underground subsurface water currents, where they could get eaten by other organisms which causes their oil to be released or where they then die from lack of sunlight and release their oil and transport it in the current to different locations to where they originally made it. Not in a process which takes millions of years of sediment formation, but in a process which could occur in process which are much faster taking only hundreds or even tens of years.

In any event, the oil making capacity of diatomic algae is well recognized by industry, with ExxonMobil taking the lead in this area with their significant studies into direct algae based biofuels as can be seen from this Headline taken from the Financial Times (2017). "ExxonMobil biofuel partnership makes oil from algae 'breakthrough' "

"*Scientists at Synthetic Genomics, the biotech company founded by genomics pioneer Craig Venter, used advanced genetic engineering to double the oil content of their algal strain from 20 to 40 per cent, without inhibiting its growth.*" *Algae can in principle produce seven times more energy per acre than corn-based ethanol, the main source of biofuel today. Other advantages include the ability of algae to grow in salt water and thrive in harsh environmental conditions—limiting the pressure on farmland and fresh water supplies. Algal oil could be processed in conventional refineries, producing fuel that can substitute directly for petrol or diesel*"

What this tells us that we do not have to be reliant on finding the naturally occurring sources of renewable oil created by diatoms, which we drill out of the ground, but we can actually farm the algae which make the oil which fills the tanks in our V8 automobiles in a more controlled and direct manner. And if algae farms, can make 7 times more energy output per acre than other sources then rather than looking to replace oil with electric technologies which is a highly inefficient process, we should instead keep the oil industry intact and simply replace the source of the oil from being the ground, but to farmed algae refineries and look to conduct research on the diatomic migratory patterns and seek to improve the naturally occurring processes to best suit our needs.

Also, as most of our oceans are filled with these oil making diatoms, it is not such of a stretch of the imagination to think that our sub-surface oceans could also be filled with these diatoms and that they could use them for either migration, or just get trapped in them which gets them moved to other locations of the world.

If you are unaware of sub-surface oceans, this particular study may be of interest for you.

"Underground ocean discovered below the Earth's surface."

"The underground ocean consists of water trapped within

minerals, and it may be the size of all of the Earth's oceans put together."

"Those are the findings of two parallel studies conducted by researchers at Florida State University, the University of Edinburgh, and Northwestern University."

"The first study, carried out by Florida State and Edinburgh, found that water can exist far deeper in the Earth's mantle than previously thought. They focused on how the water could exist so deep within the planet, and found that a mineral called brucite stores it. Their initial studies should that the water stored in the Earth's mantle could account for as much as 1.5 percent of the Earth's weight." [38]

How many of these sub-surface oceans are there? Are they interconnected? How do creatures, microbes, bacteria, diatoms and other critters make use of these oceans? Are these sub-surface oceans potential sources of new water, which could cause oceans to rise?

Do these oceans get released by volcanic eruptions? Is this water manufactured by the Earth's core? The number of questions which these sub-surface oceans raise is endless and presently we don't know the answers. We know that diatoms make oil and get drawn underground, we could therefore summise that if these sub-surface oceans are connected that they could act as migration pathways around the world, underground, which could offer an explanation as to how oil field arise deep underground. These underground oceans, open up a whole new area of research which is needed to help us better understand our planet and research into diatoms is required to help us make and extract oil in a more direct and sustainable manner.

Conclusion

Oil is not a fossil fuel, it comes from Diatoms which are present in

---

[38] https://www.earth.com/news/underground-ocean-beneath-earth/

most watery environments. These micro critters can make oil it quantities which far exceed normal farming methods, which means they are more efficient than not only normal biofuels but also wind and solar farms. Therefore rather than pouring subsidies into solar farms, which take up land, we should be segregating suitable coastal & sea areas specifically for the purpose of farming these oil producing beauties. Research should also be conducted in selecting and enhancing their oil producing capabilities through genetic modification research and also giving the best environments, nutrients and temperature and life cycle provisions which closely mimic their natural environment. Research should also be conducted into the sea-bed subduction currents so we can better understand how they get drawn underground and further research conducting into underground oceans so we can determine, how these underground oceans interact with the above ground oceans we are all familiar with, and if they are a potential pathway from one part of the world to the other, for creatures which can get into them.

Final Conclusion

I have clearly demonstrated that Frizzler climate science is nothing but "A BIG LIE!" in many different ways.

Starting from the way they lie about how a greenhouse works by pretending they work by radiation, when they do not they work by restricting convective cooling.

I have shown that how from a thermodynamic radiation point of view, by putting a greenhouse in space, the absorbing surface is colder than when compared with a flat plate on its own. No warming because there is no greenhouse effect.

I have shown that the greenhouse gas in a bottle experiment is nothing but a complete fraud and is deliberately designed to trick people by falsely explaining what is occurring and not showing comparisons with Argon and Chlorine gases, both of which are non-emissive gases, yet experience warming even greater than Carbon

Dioxide when heated using a mantle.

They do not tell the observer that light bulbs emit some 85%+ IR light and CO_2 absorbs this energy "DIRECTLY" rather than as a result of some back radiation effect.

I have shown that when you put Carbon Dioxide Gas in a light bulb with a single wire, the temperature of the wire reduces and its light output reduces.

When we examine Venus, we see that there is no "Runaway" greenhouse effect and that in fact it is hot because of high atmospheric pressures, thousands of volcanoes and a thin planetary crust.

If we were to postulate on the temperatures on two imaginary planets we saw that a pure Oxy planet would be hot and a pure Carbon Dioxide planet would be freezing cold. How could "Carbo", ever provide more warmth than "Oxy", if we placed a rock in the centre?

I have shown using Cubes and Cuboids that increasing the emissivity of surfaces decreases their temperatures, even when allowing for transparency and allowing IR radiation to escape.

I have also shown that the very basis for back radiation physics is flawed in that it is two dimensional and that relying on the 911 flat back method, for explaining radiation greenhouse effect is completely flawed and will result in wrongly determined temperatures. When combining the back plate model with other methods of heat transfer conduction, convection and latent heat, we saw the warming effect nullified and that weather patterns as explained by Hadley cells provides an example of how dominant these other methods are in heat transportation on Earth.

We then saw, that adding a large back plate as a more accurate representation of an atmosphere with a larger surface area to emit energy out to space and to take account of the three dimensional nature of the constituents of the atmosphere, as nothing blocks the radiation paths going back, forth, left, right or upwardly and only one part of the downward path allows for view factor warming. When using this more accurate model, we could see that the presence of the larger back plate atmosphere causes a cooling of the smaller front plate surface, a cooling which can only be increased by raising the emissivity of the atmosphere.

We know that CO_2 accounts for such a small proportion of atmospheric constituents that its effects can only provide negligible cooling compared to all the clouds and water vapour that is present in the atmosphere.

In the weight of all this totally overwhelming evidence, the complete absence of any evidence, whatsoever to the contrary, we can say with complete confidence that we are being lied to by the Left leaning, high taxation, socialist agenda educational and political establishments of the world, whom are using "Climate" & "Environment" & "Pollution" as nothing but fronts for their communist and quite ridiculous high tax idealism. And fake science as the con-artists tool to convincing the weak willed and naive that man-made climate change is real.

We the people, should rise up against the tide of lies and demand all Carbon Taxes be removed, all CO_2 control measures be repealed, all subsidies to unnecessary companies be removed, all Carbon Trading instruments and regulations be annulled and rendered obsolete and worthless. All building regulations that include stipulations to CO_2 control should be deleted. All fake educational pretend climate science reference books should be destroyed and confined to the dustbin of medieval, witchcraft history, where in 50 and 100 years' time hence onwards, pupils can look back at the "Frizzlers" and laugh at them for being as dumb as they are. Much like I do now.

We should put our efforts and resources into dealing with real life problems, like making energy cheaper, making cars cheaper, making houses cheaper, making medicines cheaper, improving access to jobs, improving the quality of the water supply, training people not to be pointless fake climate scientists, time wasting activists and fat cat taxmen lazing off the hard work of others but instead to be doctors, tradesmen and the engineers powering us towards a high population, high energy, high tech future.

Black Dragon

People should be schooled in how things work not fully paid up 3rd degree wizard scholars in nonsensical Marxist religiosity of adult phase, fairy tale fake fad frizzle fantasy.

Geraint Hughes

Bibliography and Referencing

Chapter One

Journals / Articles

Robert R Wood, Note on the theory of the greenhouse' *The London, Edinburgh and Dublin Philosophical Magazine and Journal of Science* (1909), p319 to 320,[Online]Available From https://archive.org/stream/londonedinburg6171909lond#page/318/mode/2up [Accessed 19th August 2017]

Websites

INDEPENDENT, *Don't believe the hype over climate headlines.* [Online] http://www.independent.co.uk/news/science/steve-connor-don't-believe-the-hype-over-climate-headlines-2180195.html [Accessed 3rd September 2017]

OECD, *Glossary of Statistical Terms.* [Online] https://stats.oecd.org/glossary/detail.asp?ID=1151 [Accessed 4th September 2017]

WORLDBANK, *Pricing Carbon,* [Online]
http://www.worldbank.org/en/programs/pricing-carbon#WhyCarbonPricing [Accessed 4th September 2017]

MEIN KAMPF, *Chapter 10, Cause of the Collapse* [Online]
http://www.mondopolitico.com/library/meinkampf/v1c10.htm
[Accessed 4th September 2017]

Chapter Two

DANISH SPACE RESEARCH INSTITUTE, *Danish Small Satellite Programme* [Online]
http://www.space.aau.dk/cubesat/documents/Cubesat_Thermal_Design.pdf [Accessed 4th September 2017]

Chapter Three

Table 3.1 Density of Gases,
THE ENGINEERING TOOLBOX, *Gases and Density* [Online]
http://www.engineeringtoolbox.com/gas-density-d_158.html
[Accessed 20th August 2017]
Table 3.2 Specific Heat Capacity of Gases,

THE ENGINEERING TOOLBOX, *Specific Heat and Individual Gas Constant of Gases* [Online]
http://www.engineeringtoolbox.com/specific-heat-capacity-gases-d_159.html [Accessed 20th August 2017]

Diagram 3.2 Blackbody Emission Spectrum at Different Temperatures, SUN, *Black-body Radiation.* [Online]
http://www.sun.org/encyclopedia/black-body-radiation
[Accessed 3rd November 2018]

Diagram 3.3 Light Bulb with a Vacuum & Gas
LAMPTECH, *Gas Filling Effects* [Online]

http://www.lamptech.co.uk/Documents/IN%20Atmosphere.ht m [Accessed 14th May 2018]

Diagram 3.4 Three in One Coils in a Gas
LAMPTECH, *Filament Coiling Effects* [Online]
http://www.lamptech.co.uk/Documents/IN%20Coiling.htm [Accessed 14th May 2018]

Chapter 4

Books

HANNU KARTTUNEN (2007) *Fundamental Astronomy*, Fifth Edition, New York : Pearson.

Journals

P B JAMES, M T ZUBER & R J PHILLIPS, (2013) *Crustal Thickness and support of topography on Venus* [Online] Journal of Geophysical Research, p859 to 875. Available from: http://onlinelibrary.wiley.com/doi/10.1029/2012JE004237/full [Accessed 20th August 2017]

S C SOLOMON & J W HEAD, (1982) *Mechanisms for Lithospheric Heat Transport on Venus: Implications for Tectonic Style and Volcanism* [Online] Journal of Geophysical Research, Vol 87, P9236 to 9246. Available from: http://topex.ucsd.edu/venus/papers/005_Solomon_Head_JGR_ 1982.pdf [Accessed 20th August 2017]

LARRY W ESPOSITO, M COPLEY,R ECKERT, L GATES, A.I.F. STEWART & H WORDEN. (1988) *Sulfur Dioxide at the Venus Cloud Tops, 1978-1986.* [Online] Journal of Geophysical Research, Vol 93, 20 May 1988, P 5267-5276. Available from: http://sci.esa.int/venus-express/50894-esposito-l-w-et-al-

1988/# [Accessed 10th September 2017]

Websites

NOVA CELESTIA, *Venus* [Online]
http://www.novacelestia.com/space_art_solar_system/venus.ht
ml
[Accessed 20th August 2017]

UNIVERSETODAY, *Clouds on Venus,* [Online]
https://www.universetoday.com/36871/clouds-on-venus/
[Accessed 20th August 2017]

METOFFICE, *Highs and Lows weather conditions,* [Online]
http://www.metoffice.gov.uk/learning/learn-about-the-weather/highs-and-lows/weather-conditions [Accessed 20th August 2017]

CHICAGO TRIBUNE, *Thick Crust or Thin?* [Online]
http://articles.chicagotribune.com/1992-09-06/features/9203210166_1_plate-tectonics-magellan-crust
[Accessed 10th September 2017]

CORNELL UNIVESRITY, *Venus* [Online]
http://www.astro.cornell.edu/~randerson/Inreach%20Web%20Page/inreach/venus.html [Accessed 10th September 2017]

LIVE SCIENCE, *Death Valley: 100 Years as Earths Hottest Spot* [Online]
https://www.livescience.com/38054-why-death-valley-hot.html
[Accessed 20th August 2017]

WANDER WISDOM, *Below Sea Level The Worlds Ten Lowest Points of Land* [Online]
https://wanderwisdom.com/travel-destinations/Below-Sea-Level-

Exploring-the-Worlds-Ten-Lowest-Points-of-Land [Accessed 20th August 2017]

NBC NEWS.COM, *Volcanoes on Venus may be young and hot* [Online] http://www.nbcnews.com/id/36286975/ns/technology_and_science-space/t/volcanoes-venus-may-be-young-hot/#.WYcBn9iou00 [Accessed 20th August 2017]

CRYSTAL LINKS, *Volcanoes off Planet* [Online] http://www.crystalinks.com/volcanoesplanets.html [Accessed 20th August 2017]

SPACE.COM, *Volcanoes on Venus Erupted Recently, New Study Suggests* [Online] https://www.space.com/34420-venus-volcanos-erupted-recently-hotspot-study-suggests.html [Accessed 20th August 2017]

EUROPEAN SPACE AGENCY, *Have Venusian Volcanoes been caught in the act?* [Online] http://www.esa.int/Our_Activities/Space_Science/Venus_Express/Have_Venusian_volcanoes_been_caught_in_the_act [Accessed 20th August 2017]

Images

Chapter 4 Image of Venus
NASA, *Global view of Venus produced by Magellan* [Online]
https://photojournal.jpl.nasa.gov/catalog/PIA00478
[Accessed 3rd November 2018]

Figure 4.1 Dickinson Impact Volcano Venus
NASA, *45 mile wide Dickinson Impact Crater and Lava Flows,* [Online]
https://photojournal.jpl.nasa.gov/jpeg/PIA00479.jpg [Accessed 3rd November 2018]

Figure 4.2 Imagery for Volcano
FREE PHOTOS, *CCO CREATIVE COMMONS*
https://pixabay.com/en/volcanoes-magma-lava-mountains-691939/ [Accessed 10th September 2017]

Chapter 5

Figure 5.1 Table of A & E Ratios
SOLAR MIRROR, *Absorptivity & Emissivity table 1 plus others* [Online]
http://www.solarmirror.com/fom/fom-serve/cache/43.html
[Accessed 20th August 2017]

Chapter 6

SCRIBD, *10 Radiant Heat Transfer, John Richard Thome.*[Online]
https://www.scribd.com/document/349596430/Slides-10-Radiation [Accessed 1st September 2017]

Chapter 7

METOFFICE, *Global Circulation Patterns,* [Online]
http://www.metoffice.gov.uk/learning/learn-about-the-weather/how-weather-works/global-circulation-patterns [Accessed 10th September 2017]

Chapter 8

Diagram 8.1 Diatom
BRITANNICA, *Diatom Algae,* [Online]
https://www.britannica.com/science/diatom [Accessed 19th May 2018]

Table 8.1 Comparison of Yields for Plant Types
OILGAE, *Algae oil yields.* [Online]
http://www.oilgae.com/algae/oil/yield/yield.html   [Accessed

19th May 2018]

PAULO CARTAXANA, SONIA CRUZ, CARLA GAMEIRO & MICHAEL KUHL. (2016) *Regulation of Intertidal Microphytobenthos Photosynthesis Over a Diel Emersion Period Is Strongly Affected by Diatom Migration Patterns* . [ONLINE] https://www.frontiersin.org/articles/10.3389/fmicb.2016.00872/full [Accessed 19th May 2018]

FINANCIAL TIMES, *ExxonMobil biofuel partnership makes oil from algae 'breakthrough'* [Online] https://www.ft.com/content/85bb7f54-54da-11e7-9fed-c19e2700005f [Accessed 19th May 2018]

EARTH.COM YOUR WORLD, *Underground ocean discovered below the Earths surface* [Online] https://www.earth.com/news/underground-ocean-beneath-earth/ [Accessed 2nd June 2018]

CPSIA information can be obtained
at www.ICGtesting.com
Printed in the USA
LVHW091057211119
637821LV00008B/726/P